George Darwin

On Figures of Equilibrium of Rotating Masses of Fluid

George Darwin

On Figures of Equilibrium of Rotating Masses of Fluid

ISBN/EAN: 9783337103644

Printed in Europe, USA, Canada, Australia, Japan

Cover: Foto ©berggeist007 / pixelio.de

More available books at **www.hansebooks.com**

XIII. *On Figures of Equilibrium of Rotating Masses of Fluid.*

By G. H. DARWIN, *M.A., LL.D., F.R.S., Fellow of Trinity College, and Plumian Professor in the University of Cambridge.*

Received April 28,—Read June 16, 1887.

[PLATES 22, 23.]

IN a previous paper* I remarked that there might be reason to suppose that the earliest form of a satellite might not be annular. Whether or not the present investigation does actually help us to understand the working of the nebular hypothesis, the idea there alluded to was the existence of a dumb-bell shaped figure of equilibrium, such as is shown in the figures at the end of this paper. These figures were already drawn when a paper by M. POINCARÉ appeared, in which, amongst other things, a similar conclusion was arrived at. My paper was accordingly kept back in order that an attempt might be made to apply the important principles enounced by him to this mode of treatment of the problem. The results of that attempt are, for reasons explained below, given in the Appendix.

The subject of figures of equilibrium of rotating masses of fluid is here considered from a point of view so wholly different from that of M. POINCARÉ that, notwithstanding his priority and the greater completeness of his work, it still appears worth while to present this paper.

The method of treatment here employed is simple of conception; but it is unfortunate that, to carry out the idea, a very formidable array of analysis is necessary.

In the last section a summary will be found of the principal conclusions, in which analysis is avoided.

§ 1. *Formulæ of Spherical Harmonic Analysis.*

Let there be two sets of rectangular axes, as shown in fig. 1; and let z be measured from o to O, whilst Z is measured from O to o; let $r^2 = x^2 + y^2 + z^2$, $R^2 = X^2 + Y^2 + Z^2$; and let $c = oO$.

Then

$$x + X = 0, \qquad y + Y = 0, \qquad z + Z = c \quad . \quad . \quad . \quad . \quad . \quad . \quad (1)$$

Let w_i, W_i, denote the solid zonal harmonics of degree i of the coordinates x, y, z, and X, Y, Z, respectively.

Now we shall require to express the solid zonal and certain tesseral harmonics of

* 'Phil. Trans.,' Part II., 1881, p. 534.

negative degrees with respect to the origin O as solid zonal and tesseral harmonics of positive degrees with respect to the origin o, and *vice versâ;* moreover, the results will have to be applied to a sphere of radius a with centre o, and to a sphere of radius A with centre O. This last clause is introduced in order to explain the introduction of the symbols a, A, in this place.

Fig. 1.

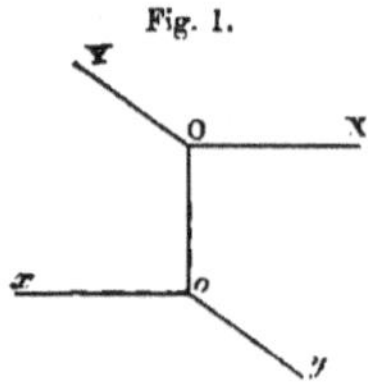

The formulæ required will be called "transference formulæ," because they are to be used in shifting the origin from one point to the other.

The obvious symmetry of our axes is such that every transference formula from O to o has its exact counterpart for transference from o to O; thus a second symmetrical formula with capital and small letters interchanged will generally be left unwritten. When necessary, θ, ϕ, will be written for co-latitude and longitude with regard to x, y, z; and Θ, Φ, for the same with respect to X, Y, Z.

Then, since

$$R^2 = r^2 + c^2 - 2rc\cos\theta,$$

we have the usual expansion in zonal harmonics

$$\frac{c}{R} = \sum_{k=0}^{k=\infty} \frac{w_k}{c^k}. \quad . \quad . \quad . \quad . \quad . \quad . \quad . \quad . \quad . \quad . \quad . \quad (2)$$

The usual formula for the derivation of the zonal harmonic of negative degree $i + 1$ from $1/R$ is

$$\frac{(-)^i}{i!}\frac{d^i}{dZ^i}\frac{1}{R} = \frac{W_i}{R^{2i+1}}. \quad . \quad . \quad . \quad . \quad . \quad (3)$$

Hence, on differentiating (2) i times with respect to Z, or, which is the same thing, with respect to $-z$, we have, from (3),

$$c\frac{W_i}{R^{2i+1}} = \frac{1}{i!}\sum_{k=0}^{k=\infty}\frac{d^i}{dz^i}\frac{w_k}{c^k}.$$

But

$$\frac{d^i}{dz^i}w_k = k(k-1)\ldots(k-i+1)\,w_{k-i} = \frac{k!}{k-i!}w_{k-i}.$$

Hence

$$c^i\frac{W_i}{R^{2i+1}} = \frac{1}{c}\sum_{k=0}^{k=\infty}\frac{k!}{i!\,k-i!}\frac{w_{k-i}}{c^{k-i}}.$$

In interpreting this formula, it will be noted that, if i is less than k, the term vanishes: hence the summation runs from $k=\infty$ to $k=i$; it is therefore better to write $k+i$ for k, and we thus obtain

$$\frac{c^i W_i}{R^{2i+1}} = \frac{1}{c}\sum_{k=0}^{k=\infty} \frac{\overline{k+i}!}{i!\,k!}\left(\frac{a}{c}\right)^k \frac{w_k}{a^k}. \quad \ldots \quad (4)$$

This is the first transference formula by which the solid zonal harmonic of degree $-i-1$ with respect to O is expressed as a series of solid harmonics of positive degree with respect to o. The formula (4) includes (2) as the particular case where $i=0$. The right-hand side of (4) is convergent for r less than a. A similar formula, convergent for r greater than a, is easily obtainable, but with this we shall not concern ourselves.

It remains to find the transference formula for certain tesseral harmonics.

If we put

$$\rho = \tfrac{1}{4}(x^2+y^2), \quad \ldots \quad (5)$$

the general expression for the zonal harmonic is

$$w_i = \Sigma(-)^k \frac{i!}{k!^2 . \overline{i-2k}!} z^{i-2k}\rho^k, \quad \ldots \quad (6)$$

where the summation extends from $k=0$ to $k=\tfrac{1}{2}i$ or $\tfrac{1}{2}(i-1)$.

From (6) we have

$$\frac{dw_i}{d\rho} = \Sigma(-)^k \frac{i!}{k!^2 . \overline{i-2k}!} k z^{i-2k}\rho^{k-1}. \quad \ldots \quad (7)$$

Now, since $r^2 = z^2 + 4\rho$, we have

$$r^2\frac{dw_i}{d\rho} = \Sigma(-)^k i!\left[\frac{-(k+1)}{\overline{k+1}!^2 . \overline{i-2k-2}!} + \frac{4k}{k!^2 . \overline{i-2k}!}\right] z^{i-2k}\rho^k$$

$$= \Sigma(-)^k i! \frac{[4k(k+1) - (i-2k)(i-2k-1)]}{(k+1) . k!^2 . \overline{i-2k}!} z^{i-2k}\rho^k. \quad \ldots \quad (8)$$

Also

$$2(2i+1)w_i = \Sigma(-)^k i! \frac{2(2i+1)(k+1)}{(k+1) . k!^2 . \overline{i-2k}!} z^{i-2k}\rho^k. \quad \ldots \quad (9)$$

Subtracting (9) from (8), and simplifying the difference, we have

$$r^2\frac{dw_i}{d\rho} - 2(2i+1)w_i = \Sigma(-)^{k+1} \frac{(i+1)(i+2) . i!}{(k+1) . k!^2 . \overline{i-2k}!} z^{i-2k}\rho^k$$

$$= \Sigma(-)^{k+1} \frac{\overline{i+2}!}{\overline{k+1}!^2 . \overline{i+2-2k-2}!}(k+1) z^{i+2-2k-2}\rho^{k+1-1}$$

$$= \frac{d}{d\rho} w_{i+2}; \quad \ldots \quad (10)$$

the last transformation being derived from (7) with $i+2$ in place of i, and $k+1$ in place of k.

Differentiate (10) with respect to ρ, and notice that $dr^2/d\rho = 4$, and we have

$$r^2 \frac{d^2 w_i}{d\rho^2} - 2(2i-1)\frac{dw_i}{d\rho} = \frac{d^2 w_{i+2}}{d\rho^2}.$$

Then, with $i+2$ in place of i,

$$r^2 \frac{d^2 w_{i+2}}{d\rho^2} - 2(2i+3)\frac{dw_{i+2}}{d\rho} = \frac{d^2 w_{i+4}}{d\rho^2}. \quad . \quad . \quad (11)$$

Now

$$\frac{d}{d\rho}\frac{w_i}{r^{2i+1}} = \frac{1}{r^{2i+3}}\left\{r^2 \frac{dw_i}{d\rho} - 2(2i+1)w_i\right\}$$

$$= \frac{1}{r^{2i+3}}\frac{dw_{i+2}}{d\rho} \text{ by (10).}$$

Differentiating again,

$$\frac{d^2}{d\rho^2}\frac{w_i}{r^{2i+1}} = \frac{1}{r^{2i+5}}\left\{r^2 \frac{d^2 w_{i+2}}{d\rho^2} - 2(2i+3)\frac{dw_{i+2}}{d\rho}\right\}$$

$$= \frac{1}{r^{2i+5}}\frac{d^2}{d\rho^2} w_{i+4} \text{ by (11),}$$

or

$$\frac{1}{r^{2i+1}}\frac{d^2 w_{i+2}}{d\rho^2} = \frac{d^2}{d\rho^2}\frac{w_{i-2}}{r^{2i-3}}. \quad . \quad . \quad . \quad (12)$$

But since $\rho = \frac{1}{4}(x^2+y^2)$, it follows that, in operating on a function involving x and y only in the form x^2+y^2,

$$\frac{d}{dx} = \tfrac{1}{2}x\frac{d}{d\rho}, \quad \frac{d}{dy} = \tfrac{1}{2}y\frac{d}{d\rho}, \quad \text{and} \quad x\frac{d}{dx} - y\frac{d}{dy} = \tfrac{1}{2}(x^2-y^2)\frac{d}{d\rho}.$$

Also

$$\frac{d^2}{dx^2} = \tfrac{1}{2}\frac{d}{d\rho} + \tfrac{1}{4}x^2\frac{d^2}{d\rho^2}, \quad \frac{d^2}{dy^2} = \tfrac{1}{2}\frac{d}{d\rho} + \tfrac{1}{4}y^2\frac{d^2}{d\rho^2},$$

so that

$$\frac{d^2}{dx^2} - \frac{d^2}{dy^2} = \tfrac{1}{4}(x^2-y^2)\frac{d^2}{d\rho^2}.$$

Now let us put

$$\delta^2 = \frac{d^2}{dx^2} - \frac{d^2}{dy^2}. \quad . \quad . \quad . \quad . \quad . \quad (13)$$

Then

$$\delta^2 = \tfrac{1}{4}(x^2-y^2)\frac{d^2}{d\rho^2},$$

and therefore (12) may be written

$$\frac{1}{r^{2i+1}}\delta^2 w_{i+2} = \delta^2 \frac{w_{i-2}}{r^{2i-3}}. \quad (14)$$

These expressions in (14) are obviously solid tesseral harmonics.

The transference formula required is for $\frac{\delta^2 W_{i+2}}{R^{2i+1}}$.

By the formula (4) we have

$$\frac{c^{i-2} W_{i-2}}{R^{2i-3}} = \frac{1}{c} \sum_{k=0}^{k=\infty} \frac{k+i-2!}{i-2!\,k!} \frac{w_k}{c^k};$$

operating on both sides by δ^2, and applying (14), we have

$$\frac{c^{i-2}}{R^{2i+1}} \delta^2 W_{i+2} = \frac{1}{c} \sum_{k=0}^{k=\infty} \frac{k+i-2!}{i-2!\,k!} \frac{\delta^2 w_k}{c^k}. \quad . \quad (15)$$

Now the general formula (6) for the zonal harmonic shows us that $d^2 w_k/d\rho^2$ is zero when $k = 0, 1, 2, 3$, and hence $\delta^2 w_k$ vanishes for the same values of k. Thus the summation in (15) is from $k = \infty$ to $k = 4$, or, if we write $k + 2$ for k, from ∞ to 2. Hence (15) gives

$$\frac{c^i}{R^{2i+1}} \delta^2 W_{i+2} = \frac{1}{c} \sum_{k=2}^{k=\infty} \frac{k+i!}{i-2!\,k+2!} \left(\frac{a}{c}\right)^k \frac{\delta^2 w_{k+2}}{a^k}. \quad . \quad . \quad (16)$$

This is the second transference formula required.

We observe that the transference of a negative zonal harmonic gives us positive zonals, and that tesseral harmonics of the type $\delta^2 W_{i+2}/R^{2i+1}$ give us harmonics of the type $\delta^2 w_{k+2}$.

§ 2. *The Mutual Influence of two Spheres of Fluid without Rotation.*

Imagine two approximately spherical masses of fluid of unit density, with their centres at the origins o and O respectively, and with mean radii a and A respectively.

We shall find that each exercises on the other certain forces, one part of which has a solid zonal harmonic of the first degree as potential. This part of the force must remain essentially unbalanced in the supposed system, but we shall see hereafter that it is balanced by the rotation to be afterwards imposed on the system.

Meanwhile it will be supposed that it is annulled in some way, and we shall content ourselves with finding the mutual influence of the spheroids, and the outstanding term of the first degree of harmonics.

Let us assume that the equations, referred to our two origins, of the surfaces of the two spheroids, when they mutually perturb one another, are

$$\left.\begin{aligned} \frac{r}{a} &= 1 + \left(\frac{A}{a}\right)^3 \sum_{i=2}^{i=\infty} \frac{2i+1}{2i-2} \left(\frac{a}{c}\right)^{i+1} h_i r^{-i} w_i \\ \frac{R}{A} &= 1 + \left(\frac{a}{A}\right)^3 \sum_{i=2}^{i=\infty} \frac{2i+1}{2i-2} \left(\frac{A}{c}\right)^{i+1} H_i R^{-i} W_{-i} \end{aligned}\right\}. \quad . \quad . \quad . \quad (17)$$

The h's and H's are unknown coefficients, to be determined.

We have now to find the potential at any point in space.

The mass of the spheroid o is $\frac{4}{3}\pi a^3$, and its potential is $\frac{4}{3}\pi a^3/r$.

The potential due to the departure from sphericity, represented by the term in h_i in the first of (17), is

$$\frac{4\pi A^3}{3c}\frac{3h_i}{2i-2}\left(\frac{a}{c}\right)^i\left(\frac{a}{r}\right)^{i+1}\frac{w_i}{r^i}. \qquad (18)$$

This is written in a form convenient for passing to the case of $r = a$. It may also be written in the form

$$\tfrac{4}{3}\pi A^3\left(\frac{a}{c}\right)^{2i+1}\frac{3h_i}{2i-2}\frac{c^i w_i}{r^{2i+1}}, \qquad (19)$$

when it is in a suitable form for application of the transference formula (4).

We shall now introduce two new symbols, namely,

$$\gamma = \left(\frac{a}{c}\right)^2, \qquad \Gamma = \left(\frac{A}{c}\right)^2. \qquad (20)$$

Then (19) may be written

$$\tfrac{4}{3}\pi A^3\left(\frac{a}{c}\right)^3\frac{3h_i\gamma^{i-1}}{2i-2}\frac{c^i w_i}{r^{2i+1}};$$

and, of course, the similar potential with the other origin is

$$\tfrac{4}{3}\pi a^3\left(\frac{A}{c}\right)^3\frac{3H_i\Gamma^{i-1}}{2i-2}\frac{c^i W_i}{R^{2i+1}}. \qquad (21)$$

The whole potential at any point of space consists of the potentials of the two spheres and of the inequalities on each. The potential of the inequalities of the sphere o may be written in the form (18), and of sphere O in the form (21).

Thus the whole potential is

$$\tfrac{4}{3}\pi a^2\cdot\frac{a}{r} + \frac{4\pi A^3}{3c}\sum_{k=2}^{k=\infty}\frac{3h_k}{2k-2}\left(\frac{a}{c}\right)^k\left(\frac{a}{r}\right)^{k+1}\frac{w_k}{r^k} \qquad (22\text{–i.})$$

$$+\frac{4\pi A^3}{3c}\cdot\frac{c}{R} + \frac{4\pi a^3}{3}\left(\frac{A}{c}\right)^3\sum_{i=2}^{i=\infty}\frac{3H_i\Gamma^{i-1}}{2i-2}\frac{c^i W_i}{R^{2i+1}}. \qquad (22\text{–ii.})$$

The first line of (22) refers to origin o, the second to origin O, and to this latter half the transference formula (4) must be applied.

Now apply (2) to the first term of the second line, and (4) to one term of the series in the second term, and we have

$$\frac{4\pi A^3}{3c}\frac{c}{R} = \frac{4\pi A^3}{3c}\sum_{k=0}^{k=\infty}\left(\frac{a}{c}\right)^k\frac{w_k}{a^k},$$

and

$$\left[\frac{4\pi a^3}{3}\left(\frac{A}{c}\right)^3\frac{3H_i\Gamma^{i-1}}{2i-2}\right]\frac{c^i W_i}{R^{2i+1}} = \frac{4\pi A^3}{3c}\left(\frac{a}{c}\right)^3\frac{3H_i\Gamma^{i-1}}{2i-2}\sum_{k=0}^{k=\infty}\frac{k+i\,!}{k\,!\,i\,!}\left(\frac{a}{c}\right)^k\frac{w_k}{a^k}.$$

Thus the second line of (22) when transferred is

$$\frac{4\pi A^3}{3c}\left[\sum_{k=0}^{k=\infty}\left(\frac{a}{c}\right)^k\frac{w_k}{a^k}+\tfrac{3}{2}\left(\frac{a}{c}\right)^3\sum_{i=2}^{i=\infty}\sum_{k=0}^{k=\infty}\frac{k+i\,!}{k\,!\,i\,!}\frac{\Gamma^{i-1}}{i-1}H_i\left(\frac{a}{c}\right)^k\frac{w_k}{a^k}\right]. \quad . \quad (22\text{–ii.})$$

Then (22–i.) and (22–ii.) together constitute the potential now entirely referred to origin o.

We want to choose the h's and H's so that each spheroid may be a level surface, save as to the outstanding term of the first degree.

In order that (17) may be a level surface, when we substitute for r its value (17) in (22), the whole potential must be constant. In effecting this substitution, we may put $r = a$ in the small terms, but in $\frac{4}{3}\pi a^3/r$ we must give it the full value (17).

The constancy of the potential is secured by making the coefficient of each harmonic term vanish separately—excepting the first harmonic, which remains outstanding by supposition.

We may consider each harmonic term by itself.

As far as concerns the term involving w_k, we have, from (22–i.) and (22–ii.), as the value of the potential,

$$\tfrac{4}{3}\pi a^2\frac{a}{r}+\frac{4\pi A^3}{3c}\left[\frac{3h_k}{2k-2}\left(\frac{a}{c}\right)^k\left(\frac{a}{r}\right)^{k+1}\frac{w_k}{r^k}+\left(\frac{a}{c}\right)^k\frac{w_k}{a^k}+\tfrac{3}{2}\left(\frac{a}{c}\right)^3\left(\frac{a}{c}\right)^k\frac{w_k}{c^k}\sum_{i=2}^{i=\infty}\frac{k+i\,!}{k\,!\,i\,!}\frac{\Gamma^{i-1}}{i-1}H_i\right],$$

and the value of r which must make this constant is

$$\frac{r}{a}=1+\left(\frac{A}{a}\right)^3\frac{2k+1}{2k-2}\left(\frac{a}{c}\right)^{k+1}h_k\frac{w_k}{r^k},$$

but in the small terms inside [ ] we may put $r = a$ simply.

Make, therefore, the substitution, and equate the coefficient of w_k to zero. On dividing that coefficient by $\frac{4\pi A^3}{3c}\cdot\left(\frac{a}{c}\right)^k$, we find

$$-\frac{2k+1}{2k-2}h_k+\frac{3h_k}{2k-2}+1+\tfrac{3}{2}\left(\frac{a}{c}\right)^3\sum_{i=2}^{i=\infty}\frac{k+i\,!}{k\,!\,i\,!}\frac{\Gamma^{i-1}}{i-1}H_i=0.$$

Therefore

$$h_k=1+\tfrac{3}{2}\left(\frac{a}{c}\right)^3\sum_{i=2}^{i=\infty}\frac{k+i\,!}{k\,!\,i\,!}\frac{\Gamma^{i-1}}{i-1}H_i, \quad . \quad . \quad . \quad . \quad . \quad . \quad (23)$$

and, by symmetry,

$$H_r=1+\tfrac{3}{2}\left(\frac{A}{c}\right)^3\sum_{i=2}^{i=\infty}\frac{r+i\,!}{r\,!\,i\,!}\frac{\gamma^{i-1}}{i-1}h_i. \quad . \quad . \quad . \quad . \quad . \quad . \quad (24)$$

Multiplying both sides of (24) by the coefficient of H_r in (23), we have

$$\tfrac{3}{2}\left(\frac{a}{c}\right)^3\frac{k+r\,!}{k\,!\,r\,!}\frac{\Gamma^{r-1}}{r-1}H_r=\tfrac{3}{2}\left(\frac{a}{c}\right)^3\frac{k+r\,!}{k\,!\,r\,!}\frac{\Gamma^{r-1}}{r-1}+(\tfrac{3}{2})^2\left(\frac{a}{c}\right)^3\left(\frac{A}{c}\right)^3\sum_{i=2}^{i=\infty}\frac{r+i\,!\,r+k\,!}{r\,!\,i\,!\,k\,!\,r\,!}\frac{\gamma^{i-1}}{i-1}\frac{\Gamma^{r-1}}{r-1}h_i.$$

Performing $\sum\limits_{r=2}^{r=\infty}$ on both sides, and substituting from (23),

$$h_k - 1 = \tfrac{3}{2}\left(\frac{a}{c}\right)^3 \sum_{r=2}^{r=\infty} \frac{k+r!}{k!\,r!}\,\frac{\Gamma^{r-1}}{r-1} + (\tfrac{3}{2})^2\left(\frac{a}{c}\right)^3\left(\frac{A}{c}\right)^3 \sum_{r=2}^{r=\infty}\sum_{i=2}^{i=\infty} \frac{r+i!\,r+k!}{r!\,i!\,k!\,r!}\,\frac{\gamma^{i-1}}{i-1}\,\frac{\Gamma^{r-1}}{r-1}\,h_i. \quad (25)$$

Now let

$$\left.\begin{aligned} (k, \Gamma) &= \sum_{r=2}^{r=\infty} \frac{k+r!}{k!\,r!}\,\frac{\Gamma^{r-1}}{r-1} \\ [k, i, \Gamma] &= \sum_{r=2}^{r=\infty} \frac{r+i!\,r+k!}{r!\,i!\,k!\,r!}\,\frac{\Gamma^{r-1}}{r-1} \end{aligned}\right\} \quad . \; . \; . \; . \; . \; . \; . \; . \; . \quad (26)$$

And (25) may be written

$$h_k = 1 + \tfrac{3}{2}\left(\frac{a}{c}\right)^3 (k, \Gamma) + (\tfrac{3}{2})^2\left(\frac{a}{c}\right)^3\left(\frac{A}{c}\right)^3 \sum_{i=2}^{i=\infty} [k, i, \Gamma]\,\frac{\gamma^{i-1}}{i-1}\,h_i. \quad . \; . \quad (27)$$

By imparting to k all integral values from 2 upwards, we get a system of linear equations for the determination of the h's, and it will appear below that as many of them may be found numerically as may be desired.

We now have to consider the series (26).

Let

$$\beta = \frac{\gamma}{1-\gamma}, \qquad B = \frac{\Gamma}{1-\Gamma},$$

and denote the operations

$$\frac{1}{\lambda!}\,\frac{d^\lambda}{d\gamma^\lambda}\,\gamma^\lambda \quad \text{or} \quad \frac{1}{\lambda!}\,\frac{d^\lambda}{d\Gamma^\lambda}\,\Gamma^\lambda \quad \text{by} \quad E^\lambda.$$

Consider the function $\gamma^{-1}E^\lambda . \gamma \log(1+\beta)$.

Now

$$\log(1+\beta) = -\log(1-\gamma) = \sum_{r=2}^{r=\infty} \frac{\gamma^{r-1}}{r-1}.$$

Therefore

$$E^k . \gamma \log(1+\beta) = \frac{1}{k!}\,\frac{d^k}{d\gamma^k} \sum_{r=2}^{r=\infty} \frac{\gamma^{k+r}}{r-1} = \sum_{r=2}^{r=\infty} \frac{k+r!}{k!\,r!}\,\frac{\gamma^r}{r-1}.$$

Thus

$$(k, \gamma) = \frac{1}{\gamma} E^k . \gamma \log(1+\beta), \qquad (k, \Gamma) = \frac{1}{\Gamma} E^k . \Gamma \log(1+B) \, . \; . \; . \quad (28)$$

Next consider the function $\gamma^{-1}E^kE^i . \gamma \log(1+\beta)$.

As before,

$$E^i . \gamma \log(1+\beta) = \sum_{r=2}^{r=\infty} \frac{i+r!}{i!\,r!}\,\frac{\gamma^r}{r-1},$$

and

$$\begin{aligned} E^kE^i . \gamma \log(1+\beta) &= \frac{1}{k!}\,\frac{d^k}{d\gamma^k} \sum_{r=2}^{r=\infty} \frac{i+r!}{i!\,r!}\,\frac{\gamma^{k+r}}{r-1}, \\ &= \sum_{r=2}^{r=\infty} \frac{i+r!\,k+r!}{i!\,r!\,r!\,k!}\,\frac{\gamma^r}{r-1}. \end{aligned}$$

Hence

$$[k, i, \gamma] = \frac{1}{\gamma} E^k E^i . \gamma \log (1 + \beta), \qquad [k, i, \Gamma] = \frac{1}{\Gamma} E^k E^i . \Gamma \log (1 + B). \quad (29)$$

We must now develop the symbolical sums of the series in (28) and (29).

The following theorems are obvious :—

$$\frac{d^n}{d\gamma^n} \gamma^p = \frac{p!}{p-n!} \gamma^{p-n}, \qquad \frac{d^n}{d\gamma^n} \log (1+\beta) = \frac{n-1!}{(1-\gamma)^n},$$

$$\frac{d^n}{d\gamma^n} (1-\gamma)^{-p} = \frac{p+n-1!}{p-1!} (1-\gamma)^{-p-n}.$$

Then, by their aid, we have from LEIBNITZ's theorem—

$$\frac{d^k}{d\gamma^k} \gamma^{k+1} \log (1+\beta) = \sum_{t=0}^{t=k} \frac{k!}{t!\, k-t!} \frac{d^t}{d\gamma^t} \gamma^{k+1} \frac{d^{k-t}}{d\gamma^{k-t}} \log (1+\beta),$$

$$= \sum_{t=0}^{t=k} \frac{k!}{t!\, k-t!} \frac{k+1!\, k-t-1!}{k-t+1!} \frac{\gamma^{k-t+1}}{(1-\gamma)^{k-t}},$$

in which we interpret $(-1)!/(1-\gamma)^0$ as $\log (1+\beta)$.

Thus

$$(k, \gamma) = \sum_{t=0}^{t=k} \frac{1}{(k-t+1)(k-t)} \frac{k+1!}{t!\, k-t!} \beta^{k-t}, \quad . \quad . \quad . \quad . \quad . \quad (30)$$

with $\beta^0/0 = \log (1+\beta)$.

Again

$$\frac{1}{i!\, k!} \frac{d^i}{d\gamma^i} \gamma^i \frac{d^k}{d\gamma^k} (\gamma^{k+1} \log (1+\beta))$$

$$= \frac{1}{i!} \sum_{t=0}^{t=k} \frac{k+1!\, k-t-1!}{t!\, k-t!\, k-t+1!} \frac{d^i}{d\gamma^i} \frac{\gamma^{i+k-t+1}}{(1-\gamma)^{k-t}},$$

$$= \gamma \sum_{t=0}^{t=k} \sum_{r=0}^{r=i} \frac{k+1!\, k-t-1!\, i+k-t+1!\, i+k-r-t-1!}{t!\, k-t!\, r!\, i-r!\, k-t+1!\, k-t-1!\, i+k-r-t+1!} \beta^{i+k-r-t}.$$

Hence

$$[k, i, \gamma]$$

$$= \sum_{t=0}^{t=k} \sum_{r=0}^{r=i} \frac{1}{(i+k-r-t+1)(i+k-r-t)} \frac{k+1!\, i+k-t+1!}{k-t+1!} \cdot \frac{\beta^{i+k-r-t}}{t!\, k-t!\, r!\, i-r!} . \quad (31)$$

In (30) and (31) the infinite series are replaced by finite series.

From the form of the series it is obvious that the result must be symmetrical with respect to k and i, so that $[k, i, \gamma] = [i, k, \gamma]$, but this is not obvious on the face of the formula (31).

We shall find, therefore, the symmetrical form of (31) for the first few terms.

If $t = k$, $r = i$, we obviously have

First term $= (k+1)(i+1) \log (1+\beta)$.

The second term arises from $t = k$, $r = i - 1$, and $t = k - 1$, $r = i$. The two corresponding values of (31) will be found to add together, and we get

Second term $= \frac{1}{4}(k+1)(i+1)[2(i+k) + ik]\beta$.

The third term arises from $t = k$, $r = i - 2$; $t = k - 1$, $r = i - 1$; $t = k - 2$, $r = i$, and we find—

$$\textit{Third term} = \frac{1}{3.2}(k+1)(i+1)\left\{\frac{i(i-1)+k(k-1)}{2!} + \frac{ik(i+2)(k+2)}{2!\,3!} + \frac{ik(i+k+1)}{2!^2}\right\}\beta^2.$$

A symmetrical form for further terms may be obtained by writing (31) first with i before k and then with k before i, and taking half the sum of the two results. In computing these coefficients it is a useful check to compute from both unsymmetrical forms, when the identity of results verifies the computation.

The following Tables have been computed from (30) and (31). The numbers are the coefficients of the quantities at the heads of the columns for the values of k and i written in the first column. The series (k, γ) is terminable with β^k, and the series $[k, i, \gamma]$ is terminable with β^{k+i}.

In $[k, i, \gamma]$ the coefficients have only been computed as far as β^6, so that the last which is given completely is $[2, 4, \gamma]$; however, with such values of β as we require, the series are carried far enough to give numerical results with sufficient accuracy.

TABLE of (k, γ).

	$\log(1+\beta)$	$+\beta$	$+\beta^2$	$+\beta^3$	$+\beta^4$	$+\beta^5$
$k=2$	3	3	$\frac{1}{2}$	..	..	..
$k=3$	4	6	2	$\frac{1}{3}$	..	..
$k=4$	5	10	5	$1\frac{2}{3}$	$\frac{1}{4}$	..
$k=5$	6	15	10	5	$1\frac{1}{2}$	$\frac{1}{5}$

TABLE of $[k, i, \gamma]$.

	$\log(1+\beta)$	$+\beta$	$+\beta^2$	$+\beta^3$	$+\beta^4$	$+\beta^5$	$+\beta^6$	$+\beta^7$
$k=2, i=2$	9	27	$18\frac{1}{2}$	8	$1\frac{1}{2}$	..	..	..
$k=2, i=3$	12	48	46	31	12	2	..	..
$k=2, i=4$	15	75	$92\frac{1}{2}$	85	$50\frac{1}{4}$	17	$2\frac{1}{2}$	..
$k=2, i=5$	18	108	163	190	$151\frac{1}{2}$	$77\frac{4}{5}$	23	&c.
$k=3, i=3$	16	84	108	103	63	22	$3\frac{1}{3}$	..
$k=3, i=4$	20	130	210	260	219	118	$36\frac{2}{3}$	&c.
$k=3, i=5$	24	186	362	552	594	$434\frac{4}{5}$	206	&c.
$k=4, i=4$	25	200	400	625	$687\frac{3}{4}$	514	$248\frac{1}{3}$	&c.
$k=4, i=5$	30	285	680	1285	$1750\frac{1}{2}$	1681	1110	&c.
$k=5, i=5$	36	405	1145	2585	4272	$5098\frac{4}{5}$	4345	&c.

We must now go back and determine the value of the outstanding potential of the first degree of harmonics, which will be annulled when rotation is imposed on the system. The potential is given in (22–i.) and (22–ii.); (22–i.) contributes nothing, and (22–ii.) gives us, for $k = 1$,

$$\frac{4\pi A^3}{3c}\left[\frac{a}{c} + \tfrac{3}{2}\left(\frac{a}{c}\right)^4 \sum_{i=2}^{i=\infty} \frac{i+1\,!}{1\,!\,i\,!}\,\frac{\Gamma^{i-1}}{i-1} H_i\right]\frac{w_1}{a}.$$

Thus, if we call u_1, U_1 the outstanding potential of the first degree, when referred to the two origins respectively, we have

$$\left.\begin{aligned} u_1 &= \frac{4\pi A^3}{3c}\left[1 + \tfrac{3}{2}\left(\frac{a}{c}\right)^3 \sum_{i=2}^{i=\infty} \frac{i+1}{i-1}\Gamma^{i-1} H_i\right]\frac{a}{c}\cdot\frac{w_1}{a} \\ U_1 &= \frac{4\pi a^3}{3c}\left[1 + \tfrac{3}{2}\left(\frac{A}{c}\right)^3 \sum_{i=2}^{i=\infty} \frac{i+1}{i-1}\gamma^{i-1} h_i\right]\frac{A}{c}\cdot\frac{W_1}{A} \end{aligned}\right\}. \qquad (32)$$

§ 3. *The Potential due to Rotation.*

Intermediate between the two origins o and O take a third Q, and take the axes of ξ and η parallel to those of x and y, and that of ζ identical with that of z. Let $Qo = d$, $QO = D$.

Then suppose that the system of the two spheroids is in uniform rotation about the axes of ξ with an angular velocity ω.

The potential Ω of the centrifugal forces is given by

$$\Omega = \tfrac{1}{2}\omega^2(\eta^2 + \zeta^2). \qquad (33)$$

But

$$\left.\begin{aligned} z &= \zeta + d, & Z &= D - \zeta, & d + D &= c \\ y &= \eta & Y &= -\eta & & \\ \xi &= x & X &= -\xi & & \end{aligned}\right\}. \qquad (34)$$

Hence

$$\begin{aligned} \Omega &= \tfrac{1}{2}\omega^2(y^2 + z^2 - 2zd + d^2) \\ &= \tfrac{1}{2}\omega^2[-\tfrac{1}{2}(x^2 - y^2) + \tfrac{1}{3}(z^2 - \tfrac{1}{2}x^2 - \tfrac{1}{2}y^2) + \tfrac{2}{3}(x^2 + y^2 + z^2) - 2zd + d^2]. \end{aligned}$$

Then, remembering that

$$w_2 = z^2 - \tfrac{1}{2}x^2 - \tfrac{1}{2}y^2, \quad w_1 = z,$$

and if we put

$$q_2 = x^2 - y^2, \quad Q_2 = X^2 - Y^2,$$

we have

$$\Omega = -\tfrac{1}{4}\omega^2 q_2 + \tfrac{1}{6}\omega^2 w_2 - \omega^2 d w_1 + \tfrac{1}{3}\omega^2 r^2 + \tfrac{1}{2}\omega^2 d^2. \qquad (35)$$

Similarly the rotation potential, when developed with reference to the other origin O, is

$$\Omega = -\tfrac{1}{4}\omega^2 Q_2 + \tfrac{1}{6}\omega^2 W_2 - \omega^2 D W_1 + \tfrac{1}{3}\omega^2 R^2 + \tfrac{1}{2}\omega^2 D^2. \qquad (36)$$

The last terms of (35) and (36) are constants, and the term in r^2, and that in R^2 are symmetrical about each origin, and so the corresponding forces can produce no departure from sphericity in either mass; thus these terms may be dropped. Next we have in (35) and (36) the outstanding potentials $-\omega^2 dw_1$ and $-\omega^2 DW_1$, which will be annulled by other similar terms, and so need not be considered now. We are left, therefore, with the terms in q_2 and w_2, or in Q_2 and W_2. The q_2 is a sectorial harmonic, the w_2 a zonal, and it will be convenient to treat them separately. We shall begin with the zonal term.

§ 4. *Disturbance due to the Zonal Harmonic Rotational Term.*

The potential whose effects we are to consider is $\frac{1}{6}\omega^2 w_2$ or $\frac{1}{6}\omega^2 W_2$, according to the origin which we are considering.

If an isolated spheroid of fluid of unit density be rotating with angular velocity ω, the ellipticity of the spheroid is $15\omega^2/16\pi$; therefore we put

$$\epsilon = \frac{15\omega^2}{16\pi}. \qquad (37)$$

Let us assume, for the equations to the two spheroids,

$$\left.\begin{aligned} \frac{r}{a} &= 1 + \tfrac{1}{3}\epsilon \frac{w_2}{r^2} + \left(\frac{A}{a}\right)^3 \sum_{i=2}^{i=\infty} \frac{2i+1}{2i-2}\left(\frac{a}{c}\right)^{i+1} l_i \frac{w_i}{r^i} \\ \frac{R}{A} &= 1 + \tfrac{1}{3}\epsilon \frac{W_2}{R^2} + \left(\frac{a}{A}\right)^3 \sum_{i=2}^{i=\infty} \frac{2i+1}{2i-2}\left(\frac{A}{c}\right)^{i+1} L_i \frac{W_i}{R^i} \end{aligned}\right\}, \qquad (38)$$

where l_i, L_i, are unknown coefficients which are to be determined. We now have to determine the potentials at any point of the inequalities (38) on the two spheroids.

The potential of the inequality $\frac{1}{3}\epsilon\, w_2/r^2$ in the first of (38) is

$$\tfrac{4}{5}\pi a^3 . a^2 . \tfrac{1}{3}\epsilon \frac{w_2}{r^5} = \tfrac{1}{4}\omega^2 a^2 \left(\frac{a}{r}\right)^3 \frac{w_2}{r^2}. \qquad (39)$$

The similar inequality in the second of (38) gives us

$$\tfrac{4}{5}\pi A^3 . A^2 . \tfrac{1}{3}\epsilon \frac{W_2}{R^5} = \frac{4\pi A^3}{3c} \tfrac{1}{5}\epsilon \left(\frac{A}{c}\right)^2 \cdot \frac{c^3 W_2}{R^5}. \qquad (40)$$

The term in l_k in the first of (38) gives us, as in § 2,

$$\frac{4\pi A^3}{3c} \frac{3l_k}{2k-2}\left(\frac{a}{c}\right)^k \left(\frac{a}{r}\right)^{k+1} \frac{w_k}{r^k}. \qquad (41)$$

The term in L_i in the second of (38) gives us, as in § 2,

$$\tfrac{4}{3}\pi a^3 \left(\frac{A}{c}\right)^3 \frac{3L_i\Gamma^{i-1}}{2i-2}\,\frac{c^i W_i}{R^{2i+1}}. \qquad \text{(42)}$$

The potential due to rotation is $\frac{1}{6}\omega^2 w_2$ or $\frac{1}{6}\omega^2 W_2$, being the second term of (35) or (36); this term we find it convenient to write

$$\tfrac{1}{6}\omega^2 a^2 \left(\frac{r}{a}\right)^2 \frac{w_2}{r^2}. \qquad \text{(43)}$$

The sums of the several terms (39), (40), (41), (42), and (43) are to be regarded as the potential of perturbing forces by which the spheroid a or the spheroid A is disturbed, and the arbitrary constants l and L are to be so chosen that each may be a figure of equilibrium.

We may consider the spheroid a by itself, and the solution for it will afford us the solution for the spheroid A by symmetry. In order to find the disturbance, the formulæ (40) and (42) must be transferred.

Now by (4), with $i = 2$,

$$\frac{4\pi A^3}{3c}\,\tfrac{1}{5}\epsilon\left(\frac{A}{c}\right)^2 \frac{c^3 W_2}{R^5} = \frac{4\pi A^3}{3c}\,\tfrac{1}{5}\epsilon\left(\frac{A}{c}\right)^2 \sum_{k=0}^{k=\infty} \frac{k+2!}{2!\,k!}\left(\frac{a}{c}\right)^k \frac{w_k}{a^k}. \qquad \text{(40')}$$

And again, by (4),

$$\frac{4\pi a^3}{3c}\left(\frac{A}{c}\right)^3 \frac{3L_i\Gamma^{i-1}}{2i-2}\,\frac{c^i W_i}{R^{2i+1}} = \frac{4\pi A^3}{3c}\,\tfrac{3}{2}\left(\frac{a}{c}\right)^3 \sum_{k=0}^{k=\infty} \frac{k+i!}{i!\,k!}\,\frac{\Gamma^{i-1}}{i-1}\,L_i\left(\frac{a}{c}\right)^k \frac{w_k}{a^k}. \qquad \text{(42')}$$

Then (39), (40'), the sum of (41) from $k = \infty$ to $k = 2$, the sum of (42') from $i = \infty$ to $i = 2$, and (43) together constitute the disturbing potential, all now referred to the origin o.

In order to find the disturbance of the spheroid a, we add the perturbing potential to $\frac{4}{3}\pi a^3/r$, give r its value (38) in this term, put $r = a$ in the perturbing potential, and make the whole potential constant by equating to zero the coefficient of each harmonic term.

We will begin by putting $r/a = 1 + \frac{1}{3}\epsilon\, w_2/r^2$, and considering only the perturbing potentials (39) and (43). We have then, for the coefficient of w_2/r^2,

$$-\tfrac{4}{3}\pi a^2.\tfrac{1}{3}\epsilon + \tfrac{1}{4}\omega^2 a^2 + \tfrac{1}{6}\omega^2 a^2.$$

Now, with the value of ϵ in (37),

$$-\tfrac{4}{3}\pi a^2.\,\tfrac{1}{3}\epsilon = -\tfrac{5}{12}\omega^2 a^2 \quad \text{and} \quad -\tfrac{5}{12} + \tfrac{1}{4} + \tfrac{1}{6} = 0.$$

Hence the coefficient of w_2/r^2 vanishes, and the term in ϵ in (38) has been properly chosen to satisfy the perturbing potentials (39) and (43).

Following the similar process with the remaining terms of (38), and equating to zero the coefficient of w_k, we have from (40′), (41), and (42′),

$$-\frac{2k+1}{2k-2}l_k + \frac{3l_k}{2k-2} + \tfrac{1}{5}\epsilon\left(\frac{A}{c}\right)^2\frac{k+2!}{k!\,2!} + \tfrac{3}{2}\left(\frac{a}{c}\right)^3\sum_{i=2}^{i=\infty}\frac{k+i!}{k!\,i!}\frac{\Gamma^{i-1}}{i-1}L_i = 0\,;$$

whence

$$l_k = \tfrac{1}{5}\epsilon\left(\frac{A}{c}\right)^2\frac{k+2!}{k!\,2!} + \tfrac{3}{2}\left(\frac{a}{c}\right)^3\sum_{i=2}^{i=\infty}\frac{k+i!}{k!\,i!}\frac{\Gamma^{i-1}}{i-1}L_i. \quad . \quad . \quad . \quad . \quad (44)$$

By symmetry, the condition that the spheroid A may be a level surface is

$$L_r = \tfrac{1}{5}\epsilon\left(\frac{a}{c}\right)^2\frac{r+2!}{r!\,2!} + \tfrac{3}{2}\left(\frac{A}{c}\right)^3\sum_{i=2}^{i=\infty}\frac{r+i!}{r!\,i!}\frac{\gamma^{i-1}}{i-1}l_i. \quad . \quad . \quad . \quad . \quad . \quad (45)$$

Multiply both sides of (45) by $\tfrac{3}{2}\left(\frac{a}{c}\right)^3\frac{r+k!}{r!\,k!}\frac{\Gamma^{r-1}}{r-1}$, and perform $\sum_{r=2}^{r=\infty}$ on the whole, and substitute from (44); and we have

$$l_k - \tfrac{1}{5}\epsilon\left(\frac{A}{c}\right)^2\frac{k+2!}{k!\,2!} = \tfrac{1}{5}\epsilon\left(\frac{a}{c}\right)^2\tfrac{3}{2}\left(\frac{a}{c}\right)^3\sum_{r=2}^{r=\infty}\frac{r+2!\,r+k!}{2!\,r!\,r!\,k!}\frac{\Gamma^{r-1}}{r-1}$$
$$+ (\tfrac{3}{2})^2\left(\frac{a}{c}\right)^3\left(\frac{A}{c}\right)^3\sum_{i=2}^{i=\infty}\sum_{r=2}^{r=\infty}\frac{r+i!\,r+k!}{r!\,i!\,r!\,k!}\frac{\gamma^{i-1}}{i-1}\frac{\Gamma^{r-1}}{r-1}l_i. \quad . \quad . \quad . \quad . \quad (46)$$

Introducing the notation (26) for the series involved in (46), we have

$$l_k = \tfrac{1}{5}\epsilon\left(\frac{A}{c}\right)^2\left\{\tfrac{1}{2}(k+1)(k+2) + \tfrac{3}{2}\left(\frac{a}{A}\right)^2\left(\frac{a}{c}\right)^3[k,\,2,\,\Gamma]\right\}$$
$$+ (\tfrac{3}{2})^2\left(\frac{a}{c}\right)^3\left(\frac{A}{c}\right)^3\sum_{i=2}^{i=\infty}[k,\,i,\,\Gamma]\frac{\gamma^{i-1}}{i-1}l_i. \quad . \quad . \quad . \quad . \quad (47)$$

Each value of k gives a similar equation, and there is a similar series of equations with small and large letters interchanged.

Now put

$$\left.\begin{aligned} l_k &= \tfrac{1}{10}\epsilon\left(\frac{A}{c}\right)^2(k+1)(k+2)\,\lambda_k \\ L_i &= \tfrac{1}{10}\epsilon\left(\frac{a}{c}\right)^2(k+1)(k+2)\,\Lambda_k \end{aligned}\right\}, \quad . \quad . \quad . \quad . \quad . \quad . \quad . \quad . \quad . \quad (48)$$

and (47) becomes

$$\lambda_k = 1 + \frac{3}{(k+1)(k+2)}\left(\frac{a}{A}\right)^2\left(\frac{a}{c}\right)^3[k,\,2,\,\Gamma]$$
$$+ (\tfrac{3}{2})^2\left(\frac{a}{c}\right)^3\left(\frac{A}{c}\right)^3\sum_{i=2}^{i=\infty}\frac{(i+1)(i+2)}{(k+1)(k+2)}[k,\,i,\,\Gamma]\frac{\gamma^{i-1}}{i-1}\lambda_i. \quad . \quad . \quad . \quad (49)$$

We attribute to k in (49) all values from ∞ to 2, and thus find a series of equations for the λ's. A similar series of equations holds for the Λ's.

We must now find the outstanding potential of the first degree of harmonics. No such term exists in (39), (41), (43), but it arises entirely out of (40′) and (42′). If we write v_1 for the outstanding potential, we have clearly

$$v_1 = \frac{4\pi A^3}{3c}\left\{\tfrac{1}{5}\epsilon\left(\frac{A}{c}\right)^2 \frac{3\,!}{2\,!\,1\,!}\frac{a}{c}\cdot\frac{w_1}{a} + \tfrac{3}{2}\left(\frac{a}{c}\right)^3 \sum_{i=2}^{i=\infty} \frac{\overline{i+1}\,!}{i\,!\,1\,!}\frac{\Gamma^{i-1}}{i-1}L_i\frac{a}{c}\frac{w_1}{a}\right\};$$

whence

$$v_1 = \frac{4\pi A^3}{3c}\tfrac{1}{10}\epsilon\left(\frac{a}{c}\right)^2\left\{6\left(\frac{A}{a}\right)^2 + \sum_{i=2}^{i=\infty}\frac{(i+1)^2(i+2)}{i-1}\Gamma^{i-1}\Lambda_i\right\}\frac{a}{c}\cdot\frac{w_1}{a}, \quad . \quad . \quad (50)$$

and, by symmetry,

$$V_1 = \frac{4\pi a^3}{3c}\tfrac{1}{10}\epsilon\left(\frac{A}{c}\right)^2\left\{6\left(\frac{a}{A}\right)^2 + \sum_{i=2}^{i=\infty}\frac{(i+1)^2(i+2)}{i-1}\gamma^{i-1}\lambda_i\right\}\frac{A}{c}\cdot\frac{W_1}{A}. \quad . \quad . \quad (51)$$

§ 5. *Disturbance due to the Sectorial Harmonic Rotational Term.*

In (35) and (36) we have found this term to be $-\tfrac{1}{4}\omega^2 q_2$ or $-\tfrac{1}{4}\omega^2 Q_2$.

We have already observed that, if the operation $\frac{d^2}{dx^2} - \frac{d^2}{dy^2}$ or δ^2 be performed on w_i, the result vanishes when $i = 1, 2, 3$.

Now, by (6),

$$w_4 = \sum_{k=0}^{k=2}(-)^k\frac{4\,!}{k\,!^2\,\overline{4-2k}\,!}z^{4-2k}\rho^k$$

$$= z^4 - \frac{4\,!}{1\,!^2\,2\,!}z^2\rho + \frac{4\,!}{2\,!^2\,0\,!}\rho^2.$$

Hence $\tfrac{1}{4}d^2w_4/d\rho^2 = 3$, and, since $\delta^2 w_4 = \tfrac{1}{4}(x^2 - y^2)\,d^2w_4/d\rho^2$, it follows that

$$q_2 = x^2 - y^2 = \tfrac{1}{3}\delta^2 w_4, \qquad \text{and} \qquad Q_2 = \tfrac{1}{3}\delta^2 W_4. \quad . \quad . \quad . \quad . \quad (52)$$

Hence the sectorial rotational term is $-\tfrac{1}{12}\omega^2\delta^2 w_4$ or $-\tfrac{1}{12}\omega^2\delta^2 W_4$; this potential is of the second order of sectorial harmonics.

Now, with ϵ as defined in (37), let us assume as the equations to the two surfaces,

$$\left.\begin{aligned}\frac{r}{a} &= 1 - \tfrac{1}{6}\epsilon\frac{\delta^2 w_4}{r^2} - \left(\frac{A}{a}\right)^3\sum_{i=2}^{i=\infty}\frac{2i+1}{2i-2}\left(\frac{a}{c}\right)^{i+1}m_i\frac{\delta^2 w_{i+2}}{r^i}\\ \frac{R}{A} &= 1 - \tfrac{1}{6}\epsilon\frac{\delta^2 W_4}{R^2} - \left(\frac{a}{A}\right)^3\sum_{i=2}^{i=\infty}\frac{2i+1}{2i-2}\left(\frac{A}{c}\right)^{i+1}M_i\frac{\delta^2 W_{i+2}}{R^i}\end{aligned}\right\} \quad . \quad . \quad . \quad . \quad (53)$$

We have now to determine the potentials of the inequalities on the two spheroids expressed by (53).

The potential of the inequality $-\frac{1}{10}\epsilon\,\delta^2 w_4/r^2$ in the first of (53) is

$$-\frac{4\pi a^3}{5}\cdot a^2\cdot\tfrac{1}{8}\epsilon\frac{\delta^2 w_4}{r^5} = -\tfrac{1}{8}\omega^2 a^2\left(\frac{a}{r}\right)^3\frac{\delta^2 w_4}{r^2}. \qquad (54)$$

The potential of the similar inequality in the second of (53) is

$$-\frac{4\pi A^3}{3c}\cdot\left(\frac{A}{c}\right)^2\tfrac{1}{10}\epsilon\frac{c^3\delta^2 W_4}{R^5}. \qquad (55)$$

The term in m_k in the first of (53) gives us

$$-\frac{4\pi A^3}{3c}\,\frac{3m_k}{2k-2}\left(\frac{a}{c}\right)^k\left(\frac{a}{r}\right)^{k+1}\frac{\delta^2 w_{k+2}}{r^2}. \qquad (56)$$

The term in M_i in the second of (53) gives us

$$-\frac{4\pi A^3}{3c}\left(\frac{a}{c}\right)^3\frac{3M_i}{2i-2}\,\Gamma^{i-1}\,\frac{c^i\,\delta^2 W_{i+2}}{R^{2i+1}}. \qquad (57)$$

Lastly, the sectorial term itself is

$$-\tfrac{1}{12}\omega^2 a^2\left(\frac{r}{a}\right)^2\frac{\delta^2 w_4}{r^2}. \qquad (58)$$

The sums of the several terms (54), (55), (56), (57), and (58) are to be regarded as the potential of perturbing forces by which the spheroid a, or the spheroid A, is disturbed, and the arbitrary constants m, M, are to be so chosen that they may each be figures of equilibrium. We may consider the spheroid a by itself, and the solution for it will afford the solution for the spheroid A by symmetry. In order to find the disturbance, the formulæ (55) and (57) must be transferred. For this purpose we require the second transference formulæ.

By (16), with $i=2$, we have for (55)

$$-\frac{4\pi A^3}{3c}\left(\frac{A}{c}\right)^2\tfrac{1}{10}\epsilon\frac{c^3\delta^2 W_4}{R^5} = -\frac{4\pi A^3}{3c}\tfrac{1}{10}\epsilon\left(\frac{A}{c}\right)^2\sum_{k=2}^{k=\infty}\frac{k+2!}{0!\,k+2!}\left(\frac{a}{c}\right)^k\frac{\delta^2 w_{k+2}}{a^k}. \qquad (55')$$

And by (16) we have for (57)

$$-\frac{4\pi A^3}{3c}\left(\frac{a}{c}\right)^3\frac{3M_i}{2i-2}\,\Gamma^{i-1}\,\frac{c^i\delta^2 W_{i+2}}{R^{2i+1}}$$

$$= -\frac{4\pi A^3}{3c}\,\tfrac{3}{2}\left(\frac{a}{c}\right)^3\sum_{k=2}^{k=\infty}\frac{k+i!}{i-2!\,k+2!}\,\frac{\Gamma^{i-1}}{i-1}M_i\left(\frac{a}{c}\right)^k\frac{\delta^2 w_{k+2}}{a^k}. \qquad (57')$$

Then (54), (55′), the sum of (56) from $k=\infty$ to $k=2$, the sum of (57′) from $i=\infty$ to $i=2$, and (58) together constitute the disturbing potential, all now referred to the origin o.

In order to find the disturbance of the spheroid a, we add the perturbing potential to $\frac{4}{3}\pi a^3/r$, give r its value (53) in this term, put $r = a$ in the perturbing potential, and make the whole potential constant by equating to zero the coefficients of each harmonic term.

We will begin by putting $r/a = 1 - \frac{1}{6}\epsilon\, \delta^2 w_4/r^2$, and considering only the perturbing potentials (54) and (58). We have then, for the coefficient of $\delta^2 w_4/r^2$,

$$\tfrac{4}{3}\pi a^3 \,.\, \tfrac{1}{6}\epsilon - \tfrac{1}{8}\omega^2 a^2 - \tfrac{1}{12}\omega^2 a^2.$$

Now, with the value of ϵ in (37),

$$\tfrac{4}{3}\pi a^3 \,.\, \tfrac{1}{6}\epsilon = \tfrac{5}{24}\omega^2 a^2, \quad \text{and} \quad \tfrac{5}{24} - \tfrac{1}{8} - \tfrac{1}{12} = 0.$$

Hence the coefficient of $\delta^2 w_4/r^2$ vanishes, and the term ϵ in (53) has been properly chosen to satisfy the perturbing potentials (54) and (58). Following the similar process with the remaining terms of (53), and equating to zero the coefficient of $\delta^2 w_{k+2}$, we have, from (55′), (56), (57′),

$$\frac{2k+1}{2k-2} m_k - \frac{3m_k}{2k-2} - \tfrac{1}{10}\epsilon\left(\frac{A}{c}\right)^2 - \tfrac{3}{2}\left(\frac{a}{c}\right)^3 \sum_{i=2}^{i=\infty} \frac{k+i\,!}{i-2\,!\,k+2\,!}\,\frac{\Gamma^{i-1}}{i-1} M_i = 0,$$

or

$$m_k = \tfrac{1}{10}\epsilon\left(\frac{A}{c}\right)^2 + \tfrac{3}{2}\left(\frac{a}{c}\right)^3 \sum_{i=2}^{i=\infty} \frac{k+i\,!}{i-2\,!\,k+2\,!}\,\frac{\Gamma^{i-1}}{i-1} M_i. \quad . \quad . \quad . \quad . \quad (59)$$

By symmetry the condition that the spheroid A may be a level surface is

$$M_r = \tfrac{1}{10}\epsilon\left(\frac{a}{c}\right)^2 + \tfrac{3}{2}\left(\frac{A}{c}\right)^3 \sum_{i=2}^{i=\infty} \frac{r+i\,!}{i-2\,!\,r+2\,!}\,\frac{\gamma^{i-1}}{i-1} m_i. \quad . \quad . \quad . \quad . \quad (60)$$

Multiply both sides of (60) by $\frac{3}{2}\left(\frac{a}{c}\right)^3 \frac{k+r\,!}{r-2\,!\,k+2\,!}\,\frac{\Gamma^{r-1}}{r-1}$, and perform $\sum\limits_{r=2}^{r=\infty}$ on the whole, and substitute from (59), and we have

$$m_k - \tfrac{1}{10}\epsilon\left(\frac{A}{c}\right)^2 = \tfrac{1}{10}\epsilon\left(\frac{a}{c}\right)^2 \tfrac{3}{2}\left(\frac{a}{c}\right)^3 \sum_{r=2}^{r=\infty} \frac{k+r\,!}{r-2\,!\,k+2\,!}\,\frac{\Gamma^{r-1}}{r-1}$$
$$+ (\tfrac{3}{2})^2\left(\frac{a}{c}\right)^3\left(\frac{A}{c}\right)^3 \sum_{r=2}^{r=\infty}\sum_{i=2}^{i=\infty} \frac{r+i\,!\,k+r\,!}{i-2\,!\,r+2\,!\,r-2\,!\,k+2\,!}\,\frac{\Gamma^{r-1}}{r-1}\,\frac{\gamma^{i-1}}{i-1} m_i. \quad (61)$$

Now let us write

$$\left.\begin{aligned} \{k, \Gamma\} &= \sum_{r=2}^{r=\infty} \frac{k+r\,!}{r-1\,!\,k+2\,!}\Gamma^{r-1} \\ \boxed{k, i, \Gamma} &= \sum_{r=2}^{r=\infty} \frac{r+i\,!\,k+r\,!}{i-1\,!\,r+2\,!\,r-1\,!\,k+2\,!}\Gamma^{r-1} \end{aligned}\right\}, \quad . \quad . \quad . \quad . \quad (62)$$

so that (61) may be written

$$m_k = \tfrac{1}{10}\epsilon\left(\frac{A}{c}\right)^2 + \tfrac{1}{10}\epsilon\left(\frac{a}{c}\right)^2 \tfrac{3}{2}\left(\frac{a}{c}\right)^3 \{k, \Gamma\} + (\tfrac{3}{2})^2\left(\frac{a}{c}\right)^3\left(\frac{A}{c}\right)^3 \sum_{i=2}^{i=\infty} \boxed{k, i, \Gamma}\, \gamma^{i-1} m_i. \quad (63)$$

Next put

$$m_k = \tfrac{1}{10}\epsilon\left(\frac{A}{c}\right)^2 \mu_k, \qquad M_k = \tfrac{1}{10}\epsilon\left(\frac{a}{c}\right)^2 \mathrm{M}_k, \quad . \quad . \quad . \quad . \quad . \quad . \quad . \qquad (64)$$

and (63) becomes

$$\mu_k = 1 + \tfrac{3}{2}\left(\frac{a}{c}\right)^3\left(\frac{a}{A}\right)^2\{k, \Gamma\} + (\tfrac{3}{2})^2\left(\frac{a}{c}\right)^3\left(\frac{A}{c}\right)^3 \sum_{i=2}^{i=\infty} \boxed{k, i, \Gamma}\, \gamma^{i-1}\mu_i. \quad . \quad . \qquad (65)$$

We attribute to k in (65) all values from ∞ to 2, and thus find a series of equations for the μ's. A similar series of equations holds for the M's.

We now have to sum the series (62).

Consider the function

$$\frac{1}{k+2}\left[(1+\beta)^{k+2} - 1\right].$$

$$\frac{1}{k+2}\left[(1+\beta)^{k+2} - 1\right] = \frac{1}{k+2}\left[(1-\gamma)^{-k-2} - 1\right] = \frac{1}{k+2}\left[\frac{k+2}{1!}\gamma + \frac{k+2\,.\,k+3}{2!}\gamma^2 + \ldots\right]$$

$$= \frac{1}{k+2}\sum_{r=2}^{r=\infty} \frac{k+r!}{k+1!\,r-1!}\gamma^{r-1} = \sum_{r=2}^{r=\infty} \frac{k+r!}{k+2!\,r-1!}\gamma^{r-1}.$$

Hence

$$\{k, \gamma\} = \frac{1}{k+2}\left[(1+\beta)^{k+2} - 1\right]. \quad . \quad . \quad . \quad . \quad . \quad . \quad . \qquad (66)$$

Next

$$\frac{1}{(k+2)\gamma^3} \cdot \frac{1}{i-1!}\frac{d^{i-2}}{d\gamma^{i-2}}\left\{\gamma^{i+1}\left[(1+\beta)^{k+2} - 1\right]\right\} = \frac{1}{\gamma^3 . i-1!}\frac{d^{i-2}}{d\gamma^{i-2}}\sum_{r=2}^{r=\infty}\frac{k+r!}{k+2!\,r-1!}\gamma^{i+r}$$

$$= \frac{1}{\gamma^3}\sum_{r=2}^{r=\infty}\frac{k+r!\,i+r!}{i-1!\,k+2!\,r-1!\,r+2!}\gamma^{r+2}.$$

Hence

$$\boxed{k, i, \gamma} = \frac{1}{(k+2)\gamma^3} \cdot \frac{1}{i-1!}\frac{d^{i-2}}{d\gamma^{i-2}}\left\{\gamma^{i+1}\left[(1+\beta)^{k+2} - 1\right]\right\}. \quad . \qquad (67)$$

The differential in (67) must now be evaluated. We have, by LEIBNITZ's theorem,

$$\frac{d^{i-2}}{d\gamma^{i-2}}\left\{\gamma^{i+1}\left[(1+\beta)^{k+2} - 1\right]\right\} = -\frac{d^{i-2}}{d\gamma^{i-2}}\gamma^{i+1} + \sum_{r=0}^{r=i-2}\frac{i-2!}{r!\,i-r-2!}\frac{d^r\gamma^{i+1}}{d\gamma^r}\frac{d^{i-r-2}}{d\gamma^{i-r-2}}(1-\gamma)^{-k-2}$$

$$= -\frac{i+1!}{3!}\gamma^3 + \sum_{r=0}^{r=i-2}\frac{i-2!}{r!\,i-r-2!}\frac{i+1!}{i-r+1!}\frac{i+k-r-1!}{k+1!}\frac{\gamma^{i-r+1}}{(1-\gamma)^{i+k-r}}$$

$$= -\frac{i+1!}{3!}\gamma^3 + \frac{\gamma^3}{(1-\gamma)^{k+2}}\sum_{r=0}^{r=i-2}\ldots\beta^{i-r-2}$$

Substituting in (67), we have

$$\boxed{k, i, \gamma} = \frac{i(i+1)}{6(k+2)}\left[-1 + (1+\beta)^{k+2}\sum_{r=0}^{r=i-2} 3!\,\frac{i-2!\,i+k-r-1!}{r!\,i-r-2!\,i-r+1!\,k+1!}\beta^{i-r-2}\right]. \qquad (68)$$

The following Tables, computed from (66) and (68), give the values of $\{k, \gamma\}$ and $\boxed{k, i, \gamma}$ as far as $k = 5$, and $k = 5$, $i = 5$.

Table of $\{k, \gamma\}$.

$$\{2, \gamma\} = \tfrac{1}{4}[(1 + \beta)^4 - 1].$$
$$\{3, \gamma\} = \tfrac{1}{5}[(1 + \beta)^5 - 1].$$
$$\{4, \gamma\} = \tfrac{1}{6}[(1 + \beta)^6 - 1].$$
$$\{5, \gamma\} = \tfrac{1}{7}[(1 + \beta)^7 - 1].$$

Table of $\boxed{k, i, \gamma}$.

$k=2, i=2$; $\frac{1}{4}[(1+\beta)^4-1]$.
$k=2, i=3$; $\frac{1}{2}[(1+\beta)^5-1]$.
$k=2, i=4$; $\frac{5}{6}[(1+\beta)^6-1]$.
$k=2, i=5$; $\frac{5}{4}[(1+\beta)^7-1]$.

$k=3, i=2$; $\frac{1}{5}[(1+\beta)^5-1]$.
$k=3, i=3$; $\frac{2}{5}[(1+\beta)^6(1+\frac{5}{4}\beta)-1]$.
$k=3, i=4$; $\frac{2}{3}[(1+\beta)^6(1+\frac{3}{2}\beta)-1]$.
$k=3, i=5$; $[(1+\beta)^7(1+\frac{7}{4}\beta)-1]$.

$k=4, i=2$; $\frac{1}{6}[(1+\beta)^6-1]$.
$k=4, i=3$; $\frac{1}{3}[(1+\beta)^6(1+\frac{3}{2}\beta)-1]$.
$k=4, i=4$; $\frac{5}{6}[(1+\beta)^6(1+3\beta+\frac{21}{10}\beta^2)-1]$.
$k=4, i=5$; $\frac{5}{6}[(1+\beta)^7(1+\frac{7}{2}\beta+\frac{14}{5}\beta^2)-1]$.

$k=5, i=2$; $\frac{1}{7}[(1+\beta)^7-1]$.
$k=5, i=3$; $\frac{2}{7}[(1+\beta)^7(1+\frac{7}{4}\beta)-1]$.
$k=5, i=4$; $\frac{10}{21}[(1+\beta)^7(1+\frac{7}{2}\beta+\frac{14}{5}\beta^2)-1]$.
$k=5, i=5$; $\frac{5}{7}[(1+\beta)^7(1+\frac{21}{4}\beta+\frac{42}{8}\beta^2+\frac{21}{5}\beta^3)-1]$.

§ 6. *Determination of the Angular Velocity of the System.*

The angular velocity of the system must now be determined in such a way as to annul the outstanding potential of the first degree of harmonics.

Referring to origin o, we have from (35) $-\omega^2 dw_1$ directly from the rotation potential; the remaining terms are $u_1 + v_1$, since the sectorial harmonic term does not contribute anything.

Thus, taking u_1 from (32), and v_1 from (50), we get for the potential

$$-\omega^2 dw_1 + \frac{4\pi A^3}{3c}\frac{w_1}{c}\left\{1 + \tfrac{3}{2}\left(\frac{a}{c}\right)^3 \sum_{i=2}^{i=\infty} \frac{i+1}{i-1}\Gamma^{i-1} H_i \right.$$
$$\left. + \tfrac{1}{10}\epsilon\left(\frac{a}{c}\right)^2\left[6\left(\frac{A}{a}\right)^2 + \tfrac{3}{2}\left(\frac{a}{c}\right)^3 \sum_{i=2}^{i=\infty} \frac{(i+1)^2(i+2)}{i-1}\Gamma^{i-1}\Lambda_i\right]\right\}.$$

Equating this to zero,

$$\frac{3\omega^2 c^3}{4\pi} d = A^3\left\{1 + \tfrac{3}{2}\left(\frac{a}{c}\right)^3 \sum_{i=2}^{i=\infty} \frac{i+1}{i-1}\Gamma^{i-1} H_i \right.$$
$$\left. + \tfrac{1}{10}\epsilon\left(\frac{a}{c}\right)^2\left[6\left(\frac{A}{a}\right)^2 + \tfrac{3}{2}\left(\frac{a}{c}\right)^3 \sum_{i=2}^{i=\infty} \frac{(i+1)^2(i+2)}{i-1}\Gamma^{i-1}\Lambda_i\right]\right\}. \quad (69)$$

And, by symmetry,

$$\frac{3\omega^2 c^3}{4\pi} D = a^3 \Bigg\{ 1 + \tfrac{3}{2}\left(\frac{A}{c}\right)^3 \sum_{i=2}^{i=\infty} \frac{i+1}{i-1} \gamma^{i-1} h_i + \tfrac{1}{10}\epsilon \left(\frac{A}{c}\right)^2 \left[6\left(\frac{a}{A}\right)^2 + \tfrac{3}{2}\left(\frac{A}{c}\right)^3 \sum_{i=2}^{i=\infty} \frac{(i+1)^2(i+2)}{i-1} \gamma^{i-1} \lambda_i \right] \Bigg\}. \qquad (70)$$

Add (70) to (69), note that $d + D = c$, and $\frac{1}{10}\epsilon = 3\omega^2/32\pi$, and solve for ω^2, and we have

$$\frac{3\omega^2}{4\pi} = \left[\left(\frac{A}{c}\right)^3 + \left(\frac{a}{c}\right)^3\right] \frac{1 + \tfrac{3}{2} \dfrac{A^3 a^3}{(A^3 + a^3)c^3} \displaystyle\sum_{i=2}^{i=\infty} \frac{i+1}{i-1} (\Gamma^{i-1} H_i + \gamma^{i-1} h_i)}{1 - \tfrac{3}{4}\left[\left(\dfrac{A}{c}\right)^5 + \left(\dfrac{a}{c}\right)^5\right] - \tfrac{3}{16}\left(\dfrac{A}{c}\right)^3 \left(\dfrac{a}{c}\right)^3 \displaystyle\sum_{i=2}^{i=\infty} \frac{(i+1)^2(i+2)}{i-1} (\gamma\Gamma^{i-1}\Lambda_i + \Gamma\gamma^{i-1}\lambda_i)}. \qquad (71)$$

Now let $1 + K$ denote the factor by which $(A/c)^3 + (a/c)^3$ is multiplied in (71). Then, if the two masses were particles, K would be zero, and (71) would simply be the usual formula connecting masses, mean motion, and mean distance in a circular orbit. Hence $1 + K$ is an augmenting factor by which the value of the square of the angular velocity must be multiplied if it be derived from the law of the periodic time of two particles revolving about one another. K, in fact, gives the correction to KEPLER's law for the non-sphericity of the masses.

This completes the solution of the problem, for we have determined the angular velocity in such a way as to justify the neglect of the harmonic terms of the first degree in §§ 2 and 4.

§ 7. *Solution of the Problem.*

We may now collect from the preceding paragraphs the complete solution of the problem.

In (38) and (53) we have found that there are terms in r/a as follows :—

$$\tfrac{1}{3}\epsilon \frac{w_2}{r^2} - \tfrac{1}{6}\epsilon \frac{\delta^2 w_4}{r^2}.$$

Now

$$w_2 = z^2 - \tfrac{1}{2}x^2 - \tfrac{1}{2}y^2, \qquad \text{and} \qquad \delta^2 w_4 = 3(x^2 - y^2);$$

hence

$$w_2 - \tfrac{1}{2}\delta^2 w_4 = z^2 - 2x^2 + y^2 = r^2 - 3x^2,$$

and these terms are therefore equal to $\epsilon(\frac{1}{3} - x^2/r^2)$.

We note that $\epsilon = 15\omega^2/16\pi = \frac{5}{4}\omega^2/\frac{4}{3}\pi$, and that $\omega^2/\frac{4}{3}\pi$ is the ratio generally written m in works on the figure of the Earth. Then, from (17), (38), (53), the equations to the two surfaces are

$$\left.\begin{aligned}\frac{r}{a} &= 1 + \epsilon\left(\tfrac{1}{3} - \frac{x^2}{r^2}\right) + \left(\frac{A}{a}\right)^3 \sum_{i=2}^{i=\infty} \frac{2i+1}{2i-2}\left(\frac{a}{c}\right)^{i+1}\left\{(h_i + l_i)\frac{w_i}{r^i} - m_i \frac{\delta^2 w_{i+2}}{r^i}\right\} \\ \frac{R}{A} &= 1 + \epsilon\left(\tfrac{1}{3} - \frac{X^2}{R^2}\right) + \left(\frac{a}{A}\right)^3 \sum_{i=2}^{i=\infty} \frac{2i+1}{2i-2}\left(\frac{A}{c}\right)^{i+1}\left\{(H_i + L_i)\frac{W_i}{R^i} - M_i \frac{\delta^2 W_{i+2}}{R^i}\right\}\end{aligned}\right\}. \quad (72)$$

From (27), (49), (65), we see that $h_2, h_3 \ldots h_i \ldots, \lambda_2, \lambda_3 \ldots \lambda_i \ldots, \mu_2, \mu_3 \ldots \mu_i \ldots$, are to be found by solving the equations resulting from all values of k from 2 to infinity in the following:—

$$\left.\begin{aligned}h_k - 1 &= \tfrac{3}{2}\left(\frac{a}{c}\right)^3 (k, \Gamma) + (\tfrac{3}{2})^2 \left(\frac{a}{c}\right)^3 \left(\frac{A}{c}\right)^3 \sum_{i=2}^{i=\infty} [k, i, \Gamma] \frac{\gamma^{i-1}}{i-1} h_i \\ \lambda_k - 1 &= \frac{3}{(k+1)(k+2)}\left(\frac{a}{c}\right)^3 \left(\frac{a}{A}\right)^3 [k, 2, \Gamma] \\ &\quad + \frac{1}{(k+1)(k+2)} (\tfrac{3}{2})^2 \left(\frac{a}{c}\right)^3 \left(\frac{A}{c}\right)^3 \sum_{i=2}^{i=\infty} [k, i, \Gamma] \frac{(i+1)(i+2)}{i-1} \gamma^{i-1} \lambda_i \\ \mu_k - 1 &= \tfrac{3}{2}\left(\frac{a}{c}\right)^3 \left(\frac{a}{A}\right)^2 \{k, \Gamma\} + (\tfrac{3}{2})^2 \left(\frac{a}{c}\right)^3 \left(\frac{A}{c}\right)^3 \sum_{i=2}^{i=\infty} \boxed{k, i, \Gamma}\, \gamma^{i-1} \mu_i\end{aligned}\right\}, \quad (73)$$

and symmetrical systems of equations for obtaining the H's, Λ's, and M's.

With the values found by the solution of these equations we then evaluate K by formula (71); and we have

$$\tfrac{4}{5}\epsilon = \frac{3\omega^2}{4\pi} = \left[\left(\frac{A}{c}\right)^3 + \left(\frac{a}{c}\right)^3\right](1 + K). \quad \ldots \ldots \quad (74)$$

We are now enabled to find the l's and m's by the formulæ (48) and (64), viz.,

$$\left.\begin{aligned}l_k &= \tfrac{1}{10}\epsilon (k+1)(k+2)\left(\frac{A}{c}\right)^2 \lambda_k \\ m_k &= \tfrac{1}{10}\epsilon \left(\frac{A}{c}\right)^2 \mu_k\end{aligned}\right\}, \quad \ldots \ldots \quad (75)$$

and the symmetrical forms give us the L's and M's.

Having thus evaluated all the auxiliary constants, (72) gives the solution of the problem.

It is well known that $\frac{5}{4} \times 3\omega^2/4\pi$ is the ellipticity of a single homogeneous mass of fluid rotating with angular velocity ω. Hence the first terms of (72) simply denote the ellipticity due to rotation in each of the masses, as if the other did not exist. Now the rigorous solution for the form of equilibrium of a rotating mass of fluid is an ellipsoid of revolution with eccentricity sin g, the value of g being given by the solution of

$$\frac{\omega^2}{2\pi} = \cot^3 g\,[(3 + \tan^2 g)\, g - 3 \tan g].^* \quad \ldots \ldots \quad (76)$$

* See, for example, THOMSON and TAIT's 'Natural Philosophy' (3), § 771, with $f = \tan g$.

Hence it will undoubtedly be more correct to construct the surface, of which the equation is (72), by regarding the part of r under the symbol Σ as the correction to the radius-vector of an ellipsoid of revolution with eccentricity determined by (76), where $\omega^2/2\pi$ is found from (74).

§ 8. *Examples of the Solution.*

The principal object of the preceding investigation is to trace the forms of the two masses when they approach to close proximity; we shall thus be able to determine the forms when they are on the point of coalescing into a single mass, and shall finally obtain at least an approximate figure of the single mass. For this purpose we require to push the approximation by spherical harmonic analysis as far as it will bear. We shall below endeavour to estimate the degree of departure from correctness involved by the use of this analysis. The results will, therefore, be worked out numerically for such values of c/a as bring the two masses close together, and it will appear that the largest value of c/a assumed for numerical solution is such that the surfaces cross; in this case the reality will be a single mass of a shape which it will be possible to draw with tolerable accuracy.

The computations are facilitated if, instead of assuming c to be an exact multiple of a, we take c^2 a multiple of a^2; that is to say, we shall take $1/\gamma$ as an integer, and therefore $1/\beta$ also an integer.

We shall in the first instance suppose the two masses to be equal. In the following examples, then, we have $A = a$, $\Gamma = \gamma$, $B = \beta$, and the two masses assume the same shape.

The computations will be carried through in detail in two cases, viz., when $\beta = \frac{1}{7}$, and when $\beta = \frac{1}{5}$. The results will also be given for $\beta = \frac{1}{6}$.

When $\beta = \frac{1}{7}$, $\gamma = \frac{1}{8}$, $c/a = 2{\cdot}8284$, and when $\beta = \frac{1}{5}$, $\gamma = \frac{1}{6}$, $c/a = 2{\cdot}449$. Thus the distances of the centres apart are $2\frac{4}{5}$ and $2\frac{2}{5}$ of the mean radius respectively. The numerical details of the two computations may be stated *pari passû*, and the numbers applying to $\beta = \frac{1}{5}$ will be distinguished by being printed in small type.

In the case of $\beta = \frac{1}{6}$, we have $\gamma = \frac{1}{7}$, $c/a = 2{\cdot}6458$; but only the final result is given, without intermediate details.

The first step is to compute the values of the several series by means of the Tables in §§ 2 and 5.

The numerical results are as follows.

TABLE of (k, γ).

	$\beta = \frac{1}{4}$	$\beta = \frac{1}{3}$
$k = 2$	·839	1·167
$k = 3$	1·433	2·012
$k = 4$	2·204	3·125
$k = 5$	3·163	4·536

TABLE of $\{k, \gamma\}$.

	$\beta = \frac{1}{4}$	$\beta = \frac{1}{3}$
$k = 2$	·177	·268
$k = 3$	·190	·298
$k = 4$	·205	·331
$k = 5$	·221	·369

TABLE of $[k, i, \gamma\}$.

	$\beta = \frac{1}{4}$	$\beta = \frac{1}{3}$
$k = 2,\ i = 2$	5·460	7·847
$k = 2,\ i = 3$	9·494	13·895
$k = 2,\ i = 4$	14·875	22·201
$k = 2,\ i = 5$	21·780	33·190
$k = 3,\ i = 2$	9·494	13·895
$k = 3,\ i = 3$	16·667	24·969
$k = 3,\ i - 4$	26·384	40·517
$k = 3,\ i = 5$	39·047	61·574
$k = 4,\ i = 2$	14·875	22·201
$k = 4,\ i = 3$	26·384	40·517
$k = 4,\ i = 4$	42·214	66·840
$k = 4,\ i = 5$	63·183	103·372
$k = 5,\ i = 2$	21·780	33·190
$k = 5,\ i = 3$	39·047	61·574
$k = 5,\ i = 4$	63·183	103·372
$k = 5,\ i = 5$	95·690	162·831

TABLE of $\boxed{k, i, \gamma.}$

	$\beta = \frac{1}{4}$	$\beta = \frac{1}{3}$
$k = 2,\ i = 2$	·177	·268
$k = 2,\ i = 3$	·475	·744
$k = 2,\ i = 4$	1·024	1·655
$k = 2,\ i = 5$	1·933	3·229
$k = 3,\ i = 2$	·190	·298
$k = 3,\ i = 3$	·519	·844
$k = 3,\ i = 4$	1·137	1·921
$k = 3,\ i = 5$	2·183	3·837
$k = 4,\ i = 2$	·205	·331
$k = 4,\ i = 3$	·560	·961
$k = 4,\ i = 4$	1·266	2·238
$k = 4,\ i = 5$	2·471	4·578
$k = 5,\ i = 2$	·221	·369
$k = 5,\ i = 3$	·624	1·096
$k = 5,\ i = 4$	1·412	2·616
$k = 5,\ i = 5$	2·803	5·479

With these values for the series, we have to compute the coefficients of the systems of simultaneous equations (73). The equations lend themselves more readily to solution if we consider $h_i - 1$, $\lambda_i - 1$, $\mu_i - 1$, as the unknowns instead of h_i, λ_i, μ_i. The results are given in the following equations.

The upper coefficients correspond to the case of $\beta = \frac{1}{7}$; the lower ones, printed in small type, to the case of $\beta = \frac{1}{5}$.

$h_2 - 1 = \cdot05902 + \cdot00300\,(h_2-1) + \cdot00033\,(h_3-1) + \cdot00004\,(h_4-1) + \cdot00001\,(h_5-1) + \ldots$

$\qquad\qquad\;\;\cdot13516 \qquad\cdot01362 \qquad\qquad\cdot00201 \qquad\qquad\cdot00036 \qquad\qquad\cdot00007$

$h_3 - 1 = \cdot10086 + \cdot00522 \qquad\quad + \cdot00057 \qquad\quad + \cdot00008 \qquad\quad + \cdot00001 \qquad + \ldots$

$\qquad\qquad\;\;\cdot23385 \qquad\cdot02412 \qquad\qquad\cdot00361 \qquad\qquad\cdot00065 \qquad\qquad\cdot00012$

$h_4 - 1 = \cdot15529 + \cdot00817 \qquad\quad + \cdot00091 \qquad\quad + \cdot00012 \qquad\quad + \cdot00002 \qquad + \ldots$

$\qquad\qquad\;\;\cdot36467 \qquad\cdot03854 \qquad\qquad\cdot00586 \qquad\qquad\cdot00108 \qquad\qquad\cdot00021$

$h_5 - 1 = \cdot22321 + \cdot01196 \qquad\quad + \cdot00134 \qquad\quad + \cdot00018 \qquad\quad + \cdot00003 \qquad + \ldots$

$\qquad\qquad\;\;\cdot53150 \qquad\cdot05762 \qquad\qquad\cdot00891 \qquad\qquad\cdot00165 \qquad\qquad\cdot00033$

. . . .

$\lambda_2 - 1 = \cdot06400 + \cdot00300\,(\lambda_2-1) + \cdot00054\,(\lambda_3-1) + \cdot00011\,(\lambda_4-1) + \cdot00002\,(\lambda_5-1) + \ldots$

$\qquad\qquad\;\;\cdot15158 \qquad\cdot01362 \qquad\qquad\cdot00335 \qquad\qquad\cdot00089 \qquad\qquad\cdot00023$

$\lambda_3 - 1 = \cdot06677 + \cdot00313 \qquad\quad + \cdot00057 \qquad\quad + \cdot00011 \qquad\quad + \cdot00002 \qquad + \ldots$

$\qquad\qquad\;\;\cdot16114 \qquad\cdot01447 \qquad\qquad\cdot00361 \qquad\qquad\cdot00098 \qquad\qquad\cdot00026$

$\lambda_4 - 1 = \cdot06976 + \cdot00327 \qquad\quad + \cdot00060 \qquad\quad + \cdot00012 \qquad\quad + \cdot00002 \qquad + \ldots$

$\qquad\qquad\;\;\cdot17174 \qquad\cdot01542 \qquad\qquad\cdot00391 \qquad\qquad\cdot00108 \qquad\qquad\cdot00029$

$\lambda_5 - 1 = \cdot07297 + \cdot00342 \qquad\quad + \cdot00064 \qquad\quad + \cdot00013 \qquad\quad + \cdot00003 \qquad + \ldots$

$\qquad\qquad\;\;\cdot18352 \qquad\cdot01646 \qquad\qquad\cdot00424 \qquad\qquad\cdot00118 \qquad\qquad\cdot00033$

. . . .

$\mu_2 - 1 = \cdot01184 + \cdot00010\,(\mu_2-1) + \cdot00003\,(\mu_3-1) + \cdot00001\,(\mu_4-1) + \cdot00000\,(\mu_5-1) + \ldots$

$\qquad\qquad\;\;\cdot02818 \qquad\cdot00047 \qquad\qquad\cdot00022 \qquad\qquad\cdot00008 \qquad\qquad\cdot00003$

$\mu_3 - 1 = \cdot01274 + \cdot00010 \qquad\quad + \cdot00004 \qquad\quad + \cdot00001 \qquad\quad + \cdot00000 \qquad + \ldots$

$\qquad\qquad\;\;\cdot03127 \qquad\cdot00052 \qquad\qquad\cdot00024 \qquad\qquad\cdot00009 \qquad\qquad\cdot00003$

$\mu_4 - 1 = \cdot01374 + \cdot00011 \qquad\quad + \cdot00004 \qquad\quad + \cdot00001 \qquad\quad + \cdot00000 \qquad + \ldots$

$\qquad\qquad\;\;\cdot03478 \qquad\cdot00057 \qquad\qquad\cdot00028 \qquad\qquad\cdot00011 \qquad\qquad\cdot00004$

$\mu_5 - 1 = \cdot01482 + \cdot00012 \qquad\quad + \cdot00004 \qquad\quad + \cdot00001 \qquad\quad + \cdot00000 \qquad + \ldots$

$\qquad\qquad\;\;\cdot03879 \qquad\cdot00064 \qquad\qquad\cdot00032 \qquad\qquad\cdot00013 \qquad\qquad\cdot00004$

. . . .

The solutions of these equations are obviously found by an easy approximation; they are

$$h_2 = 1\cdot0593 \qquad \lambda_2 = 1\cdot0642 \qquad \mu_2 = 1\cdot0118$$
$$1\cdot1377 \qquad 1\cdot1544 \qquad 1\cdot0282$$

$$h_3 = 1\cdot1012 \qquad \lambda_3 = 1\cdot0670 \qquad \mu_3 = 1\cdot0127$$
$$1\cdot2382 \qquad 1\cdot1642 \qquad 1\cdot0313$$

$$h_4 = 1\cdot1559 \qquad \lambda_4 = 1\cdot0700 \qquad \mu_4 = 1\cdot0137$$
$$1\cdot3719 \qquad 1\cdot1750 \qquad 1\cdot0348$$

$$h_5 = 1\cdot2241 \qquad \lambda_5 = 1\cdot0732 \qquad \mu_5 = 1\cdot0148$$
$$1\cdot5424 \qquad 1\cdot1870 \qquad 1\cdot0388$$

the small figures corresponding, as before, to the case of $\beta = \frac{1}{5}$.

With these values of the h's and λ's, I find

$$2\Sigma\frac{i+1}{i-1}\gamma^{i-1}h_i = \underset{1\cdot3005}{\cdot8718}\,; \quad 2\Sigma\frac{(i+1)^2(i-2)}{i-1}\gamma^i\lambda_i = \underset{2\cdot8542}{1\cdot3949}\,; \quad \tfrac{3}{2}\left(\frac{a}{c}\right)^5 = \underset{\cdot01701}{\cdot00829},$$

the summations, of course, stopping with $i = 5$.

Applying these in (71), we have, when

$$\beta = \tfrac{1}{7},\ 1 + K = \frac{1 + \cdot02891}{1 - \cdot00880} = 1\cdot0380; \text{ or, when } \beta = \tfrac{1}{5},\ 1 + K = \frac{1 + \cdot0664}{1 - \cdot0195} = 1\cdot0877,$$

whence

$$\tfrac{4}{5}\epsilon = \frac{3\omega^2}{4\pi} = \underset{\cdot13608}{\cdot08839} \times \underset{1\cdot0877}{1\cdot0380} = \underset{\cdot1481}{\cdot09175}.$$

Thus the angular velocity of the system has been found.

Next we have

$$\tfrac{1}{10}\epsilon\left(\frac{a}{c}\right)^2 = \underset{\cdot00309}{\cdot001434}.$$

Introducing this into (48) and (64) with the previously found values of the λ's and μ's,

$$\left.\begin{aligned} l_2 &= \underset{\cdot0428}{\cdot0183}, & m_2 &= \underset{\cdot0032}{\cdot00145}, \\ l_3 &= \underset{\cdot0720}{\cdot0306}, & m_3 &= \underset{\cdot0032}{\cdot00145}, \\ l_4 &= \underset{\cdot1089}{\cdot0460}, & m_4 &= \underset{\cdot0032}{\cdot00145}, \\ l_5 &= \underset{\cdot1541}{\cdot0646}, & m_5 &= \underset{\cdot0032}{\cdot00145}, \end{aligned}\right\}\text{; and hence}\left\{\begin{aligned} h_2 + l_2 &= \underset{1\cdot1806}{1\cdot0776}, \\ h_3 + l_3 &= \underset{1\cdot3100}{1\cdot1318}, \\ h_4 + l_4 &= \underset{1\cdot4808}{1\cdot2019}, \\ h_5 + l_5 &= \underset{1\cdot6965}{1\cdot2887}. \end{aligned}\right.$$

By taking the differences of $h + l$, we may conclude that

$$h_6 + l_6 = \underset{1\cdot96}{1\cdot39},$$

and this sixth harmonic term will now be included.

It appears from the values of the m's that the harmonics of the type $\delta^2 w_{i+2}$ are practically negligible, excepting the term $\delta^2 w_4$, and that in that we may neglect the part depending on m_2.

Now, if r denotes the radius-vector due to the rotation, and δr the increase of radius-vector due to the mutual influence of the two masses, we have

$$\frac{\delta r}{a} = \underset{\cdot2008}{\cdot1191}\frac{w_2}{r^2} + \underset{\cdot0637}{\cdot0309}\frac{w_3}{r^3} + \underset{\cdot0252}{\cdot0100}\frac{w_4}{r^4} + \underset{\cdot0108}{\cdot0035}\frac{w_5}{r^5} + \underset{\cdot0048}{\cdot0013}\frac{w_6}{r^6} + \ldots \quad . \quad . \quad . \quad (77)$$

We next have to consider the values of r, the radius-vector of the ellipsoid, due to rotation.

We might compute from the spherical harmonic formula

$$\frac{r}{a} = 1 + \epsilon\left(\frac{1}{3} - \frac{x^2}{r^2}\right).$$

The results so computed will be compared with the others computed as shown below.

The following Table of the angular velocity and corresponding eccentricity e of the equilibrium ellipsoid of revolution is extracted from THOMSON and TAIT'S 'Natural Philosophy,' §772:—

e	$\frac{\omega^2}{2\pi}$
·3	·0243
·4	·0436
·5	·0690
·6	·1007
·7	·1387
·8	·1816

From this we find by interpolation that, when $3\omega^2/4\pi =$ ·09175, $e =$ ·472; and, when $3\omega^2/4\pi =$ ·1481, $e =$ ·594.

These, then, are the eccentricities of the ellipsoids whose radius-vector is r in the two cases $\beta = \frac{1}{7}$, $\beta = \frac{1}{5}$.

The equations to the generating ellipses are

$$\frac{r}{a} = \frac{1 - \cdot 0806}{1 - \cdot 2228 \cos^2\theta} \text{ for } \beta = \tfrac{1}{7}, \qquad \text{and} \qquad \frac{r}{a} = \frac{1 - \cdot 1353}{1 - \cdot 3535 \cos^2\theta} \text{ for } \beta = \tfrac{1}{5}.$$

The following are the computed values of r/a for each 15° of θ, the latitude, the small figures written below appertaining to the case of $\beta = \frac{1}{5}$.

$\theta =$	0°	15°	30°	45°	60°	75°	90°
$\beta = \frac{1}{7} : \frac{r}{a} =$	1·0429,	1·0330,	1·0074,	·9753,	·9461,	·9264,	·9194
$\beta = \frac{1}{5}$:	1·075,	1·050,	1·009,	·953,	·900,	·875,	·865

Computing from the spherical harmonic formula, I find

$\beta = \frac{1}{7} : \frac{r}{a} =$	1·0382,	1·0305,	1·0096,	·9809,	·9522,	·9312,	·9235
$\beta = \frac{1}{5}$:	1·0616,	1·0490,	1·0154,	·9692	·9230,	·8892,	·8768

The greatest discrepancy occurs when $\beta = \frac{1}{5}$ and $\theta = 90°$, and the difference between the two results is $\frac{1}{70}$ of either. It follows, therefore, that in drawing the figures it is not of much importance which results we take. But, as above remarked, the radius-vectors computed from the true ellipsoidal figure are the more correct.

The formula (77) for δr consists of a series of zonal harmonics. The pole of symmetry of these harmonics lies in the equator of the ellipsoid of revolution defined by r, and is that point of each mass which lies nearest to the other. Then, denoting by θ co-latitude estimated from this pole, I find that the numerical values of δr for each 15° of θ are as follows :—

	$\theta =$ 0°	15°	30°	45°	60°	75°	90°
$\beta = \frac{1}{7}$: $\frac{\delta r}{a} =$	+ ·165,	+ ·141,	+ ·084,	+ ·019,	− ·031,	− ·055,	− ·056,
$\beta = \frac{1}{5}$:	+ ·280,	+ ·257,	+ ·142,	+ ·024.	− ·060,	− ·094,	− ·092,

	105°	120°	135°	150°	165°	180°
$\beta = \frac{1}{7}$	− ·037,	− ·004,	+ ·032,	+ ·065,	+ ·088,	+ ·096
$\beta = \frac{1}{5}$	− ·059,	− ·002,	+ ·055.	+ ·106,	+ ·143,	+ ·155

These have to be combined with r, so as to give the radius-vectors of the mass of fluid along two sections, one perpendicular to the axis of rotation (which may be called the equatorial section), the other through the axis and the two centres (which may be called the section through the prime meridian). Taking the case of $\beta = \frac{1}{7}$, we add the successive values of δr to the equatorial value of r, viz., 1·043, and thus find the equally-spaced radius-vectors along the equatorial section. Next we add the successive values of δr to the corresponding values of r, and thus find the equally-spaced radius-vectors along the prime meridian. The results are as follows :—

$\beta = \frac{1}{7}$:	$\theta =$	0°	15°	30°	45°	60°	75°	90°
Equator,	$\frac{r + \delta r}{a} =$	1·208,	1·184,	1·126,	1·062,	1·012,	·988,	·987,
Pr. Merid.	$=$	1·208,	1·174,	1·091,	·994,	·915,	·871,	·863,

	105°	120°	135°	150°	165°	180°
Equator	1·006,	1·039,	1·075,	1·108,	1·131,	1·139
Pr. Merid.	·889,	·942,	1·008,	1·072,	1·121,	1·139

These results apply to the case of $\beta = \frac{1}{7}$; those for $\beta = \frac{1}{5}$ are found in the same way, and are given in the figures referred to below.

When $\beta = \frac{1}{7}$ the distance between the centres is given by $c/a = 2·828$. I have also worked out the case of $\beta = \frac{1}{6}$, although none of the numerical details are given here.

In figs. 2, 3, 4, and 5 (Plates 22 and 23) are exhibited the figures which result from some of these computations.

Figs. 2 and 3 refer to $\beta = \frac{1}{5}$, 4 and 5 to that of $\beta = \frac{1}{6}$, and the numerical values for $\beta = \frac{1}{7}$, given above, make it easy to draw a figure for $\beta = \frac{1}{7}$.

Since in these cases the masses are equal, the two halves of the figure are the images of one another. The numerical value of each radius-vector is entered on the plates; and other numerical data and explanations are given.

Figs. 2 and 3 correspond to $\beta = \frac{1}{5}$, and here the figures as computed cross one another. The reality must, therefore, be two bulbs joined by a stalk, like a dumb-bell. The dotted lines have been filled in conjecturally, and must show pretty closely what that single figure, formed by the coalescence of the two masses, must be.

Figs. 4 and 5 show in a similar manner the case of $\beta = \frac{1}{6}$, and here the two masses are separate, although nearly in contact. When $\beta = \frac{1}{7}$ the shapes present similar characters, but are wider apart.

§ 9. *On the Use of Spherical Harmonic Analysis as a Method of Approximation.*

Spherical harmonic analysis gives less accuracy as the bodies considered depart more and more from spheres. How far, then, do our results present an approach to accuracy? To answer this question, we have to find how nearly the potentials at the surfaces of these figures may be computed from the spherical harmonic formulæ.

It would be laborious to make an accurate computation of the potential, and it fortunately appears to be unnecessary to do so, since a sufficient answer may be obtained in another way.

The potential of an ellipsoid of revolution may be computed either rigorously or by harmonic analysis. With a certain degree of eccentricity the approximate result will agree badly with the rigorous one.

If the ellipsoid consists of a fluid of unit density, there is a certain angular velocity which makes it a level surface. If ω be that angular velocity, then we know that the spherical harmonic solution would give $1 - 15\omega^2/16\pi$ as the ratio of the minor to the major axis. If then c, a, are the rigorous values of the minor and major semi-axes, the harmonic approximation is good if c/a does not differ much from $1 - 15\omega^2/16\pi$.

If we denote by $1 - \mu$ the factor by which the approximate value of the ratio of the axes is to be multiplied in order to obtain the rigorous value, we have

$$\mu = 1 - \frac{c/a}{1 - 15\omega^2/16\pi},$$

and μ may be regarded as a measure of inaccuracy.

A table of the values of $\omega^2/2\pi$, corresponding to various eccentricities $e = \sqrt{(1 - (c/a)^2)}$, is computed from the transcendental equation in THOMSON and TAIT's 'Natural Philosophy,' § 772. From these I compute as follows:—

$e = \sqrt{\left(1 - \frac{c^2}{a^2}\right)}$	$\frac{c}{a}$	$1 - \frac{15\omega^2}{16\pi}$	Difference.	$\frac{1}{\mu}$
·1	·9949	·9049	·0000	Large
·2	·9798	·9799	·0001	9799
·3	·9539	·9544	·0005	1909
·4	·9165	·9182	·0017	540
·5	·8660	·8705	·0045	193
·6	·8000	·8111	·0111	73
·7	·7141	·7399	·0258	29
·8	·6000	·6595	·0595	11·1
·9	·4359	·5869	·1510	3·9

The measures of inaccuracy corresponding to the values of e in the first column, or the values of c/a in the second, are the reciprocals of the numbers in the last column. We thus see that there is still a considerable degree of approximation when $e = ·8$, or when the ratio of the axes is 3 to 5, for the measure of inaccuracy is $\frac{1}{11}$; but for $e = ·9$ the approximation is bad.

Now the shapes of certain egg-like bodies have been computed by the spherical harmonic method, and it seems safe to assume that the approximation has given about the same degree of accuracy as would hold in the case of an ellipsoid of revolution whose minor axis bears to its major axis the same ratio as the shorter axis of the egg to the longer.

Turning now to our computation, and considering only the more elongated or meridional sections, we see that, when $\beta = \frac{1}{5}$, the longer axis is $1·355 + 1·230 = 2·585$, and the shorter $2(1 - ·227) = 1·546$; and the ratio $1·546 : 2·585$ is ·6, which corresponds to the measure of inaccuracy 1/11·1. It might, however, be more legitimate to adopt two different measures, and at the pointed end of the egg to take the ratio $·773 : 1·355 = ·57$, which will correspond to a measure of inaccuracy about $\frac{1}{10}$; and at the blunt end to take the ratio $·773 : 1·230 = ·63$, which would correspond to a measure of inaccuracy $\frac{1}{12}$ or $\frac{1}{13}$.

In the case of $\beta = \frac{1}{5}$ the two masses cross one another, and the result has been used to give an approximate picture of the dumb-bell figure of equilibrium. We now see that even in this case there is a sufficient degree of approximation to give a very good idea of the accurate result.

In the case of the meridional section, where $\beta = \frac{1}{6}$, we have for the ratio of axes at the pointed end of the egg $\frac{1 - ·1717}{1·2712} = \frac{·8283}{1·2712} = ·65$, and measure of inaccuracy about $\frac{1}{25}$; and at the blunt end $\frac{1 - ·1717}{1·1746} = \frac{·8283}{1·1746} = ·71$, and measure of inaccuracy $\frac{1}{29}$.

In the case of $\beta = \frac{1}{7}$ the similar figures are, for the pointed end, $\frac{1 - ·137}{1·208} = \frac{·863}{1·208} = ·72$, and measure of inaccuracy about $\frac{1}{30}$; and for the blunt end $\frac{1 - ·137}{1·139} = \frac{·863}{1·139} = ·76$, and measure of inaccuracy perhaps about $\frac{1}{50}$.

It thus appears that as the bodies recede the accuracy increases with great rapidity, and in the two cases considered last it is hardly necessary, from a physical point of view, to consider greater accuracy than that attained.

It must be remarked, however, that this way of estimating the degree of inaccuracy must necessarily give much too unfavourable a view.

If we have a single mass of fluid departing considerably from the spherical form, it is clear that the potential computed on the hypothesis of a layer of surface density on the true sphere will come to depart largely from the potential at the surface of the fluid. If, however, we compute the potential of such a mass at points a little remote from the surface, the approximation will be much closer. Now, where there are two masses, as in our problem, the potential at the surface of either mass consists of two parts, one due to the mass itself, the other due to the other mass. As regards the first of these two parts, the above criterion is applicable, but as regards the second part it gives too unfavourable a view.

Now in the case of the single mass the deformative forces due to centrifugal force are considerably vitiated by computation at the spherical surface instead of the true surface, whilst in the case of the two masses the tide-generating forces are computed with greater accuracy than is shown by the criterion.

Under these circumstances it has appeared worth while to give another figure below, which, judged by the criterion, would be no approximation at all.

The reasons for giving this figure will be stated when we come to it.

§ 10. *To find the Moment of Momentum of the System.*

Rotating figures of equilibrium are classified according to the amount of moment of momentum with which they are endued. It is, therefore, interesting to determine the moment of momentum of the systems now under consideration.

We must begin by finding the moments of inertia of the two masses. Let δI, δi, denote the moments of inertia of the shells of zero mass lying on the mean spheres of radii A, a.

Then

$$\delta i = \iint (y^2 + z^2)(r - a)\, a^2\, d\varpi,$$

where $d\varpi = \sin\theta\, d\theta\, d\phi$, and where the integral is taken throughout angular space.

Now

$$y^2 + z^2 = \tfrac{2}{3}a^2 + \tfrac{1}{3}a^2\left(\frac{w_2}{r^2} - \tfrac{1}{2}\frac{\delta^2 w_4}{r^2}\right),$$

and $r - a$ is the sum of a series of harmonics. Then, in consequence of the properties of harmonic functions, we need only consider the harmonics of the second degree in $r - a$, and

$$\delta i = \tfrac{1}{3}a^5 \iint \left\{\tfrac{1}{3}\epsilon\left[\left(\frac{w_2}{r^2}\right)^2 + \left(\tfrac{1}{2}\frac{\delta^2 w_4}{r^2}\right)^2\right] + \tfrac{5}{2}\left(\frac{A}{c}\right)^3\left[(h_2 + l_2)\left(\frac{w_2}{r^2}\right)^2 + 2m_2\left(\tfrac{1}{2}\frac{\delta^2 w_4}{r^2}\right)^2\right]\right\} d\varpi.$$

But

$$\iint \left(\frac{w_2}{r^2}\right)^2 d\varpi = \tfrac{4}{5}\pi, \qquad \iint \left(\tfrac{1}{2}\frac{\delta^2 w_4}{r^2}\right)^2 d\varpi = \tfrac{4}{5}\pi \,.\, 3, \qquad \tfrac{1}{3}\epsilon = \frac{5\omega^2}{16\pi},$$

and the moment of inertia of the mean sphere is $\frac{8}{15}\pi a^5$; hence, if we write

$$f = \frac{5\omega^2}{8\pi} + \tfrac{5}{3}\left(\frac{A}{c}\right)^3 [\tfrac{1}{2}(h_2 + l_2) + 3m_2]$$

$$F = \frac{5\omega^2}{8\pi} + \tfrac{5}{3}\left(\frac{a}{c}\right)^3 [\tfrac{1}{2}(H_2 + L_2) + 3M_2],$$

the moments of inertia, i and I, are given by

$$i = \tfrac{8}{15}\pi a^5 (1 + f),$$

$$I = \tfrac{8}{15}\pi a^5 (1 + F).$$

We already have in (71)

$$\frac{3\omega^2}{4\pi} = \left[\left(\frac{A}{c}\right)^3 + \left(\frac{a}{c}\right)^3\right](1 + K).$$

Hence the sum of the rotational momenta of the two masses is

$$(i + I)\,\omega = \tfrac{2}{5}\left(\frac{4\pi}{3}\right)^{\frac{1}{2}} [a^5(1 + f) + A^5(1 + F)]\left[\left(\frac{a}{c}\right)^3 + \left(\frac{A}{c}\right)^3\right]^{\frac{1}{2}} (1 + K)^{\frac{1}{2}}.$$

The whole system revolves orbitally about the centre of inertia with an angular velocity ω: hence the orbital momentum is

$$\tfrac{4}{3}\pi\,[a^3\omega\, d^2 + A^3\omega\, D^2].$$

But

$$d = \frac{A^3 c}{a^3 + A^3}, \qquad D = \frac{a^3 c}{a^3 + A^3}.$$

Hence the orbital momentum is

$$\frac{\frac{4}{3}\pi\, A^3 a^3}{A^3 + a^3}\,\omega c^2,$$

and this is equal to

$$(\tfrac{4}{3}\pi)^{\frac{3}{2}} \frac{a^3 A^3 c^{\frac{1}{2}}}{(A^3 + a^3)^{\frac{1}{2}}} (1 + K)^{\frac{1}{2}}.$$

It will be convenient to refer the mass to the radius of a sphere of the same mass as the sum of the two.

Let b be the radius of such a sphere; then

$$b^3 = A^3 + a^3.$$

Thus the whole moment of momentum is

$$(\tfrac{4}{3}\pi)^{\frac{3}{2}}\, b^5 \left\{\tfrac{2}{5}\left[\left(\frac{a}{b}\right)^5 (1 + f) + \left(\frac{A}{b}\right)^5 (1 + F)\right]\left(\frac{b}{c}\right)^{\frac{3}{2}} + \left(\frac{a}{b}\right)^3 \left(\frac{A}{b}\right)^3 \left(\frac{c}{b}\right)^{\frac{1}{2}}\right\} (1 + K)^{\frac{1}{2}}.$$

We shall therefore compute the coefficient of $(\frac{4}{3}\pi)^{\frac{3}{2}} b^5$.

Computing from this formula, I find the following values of the moment of momentum in the case where the masses are equal, when

$$\beta = \tfrac{1}{7}, \qquad \left(\frac{4\pi}{3}\right)^{\frac{3}{2}} b^5 \times \cdot 468$$

$$\beta = \tfrac{1}{6}, \qquad \times \cdot 472$$

$$\beta = \tfrac{1}{5}, \qquad \times \cdot 482$$

Now I find by a numerical investigation* that, if we imagine a mass of fluid equal to $\frac{4}{3}\pi b^3$ rotating in the form of a Jacobian ellipsoid of three unequal axes, then, when the momentum is $(\frac{4}{3}\pi)^{\frac{3}{2}} b^5 \times \cdot 392$, the axes of the ellipsoid are $1 \cdot 898b$, $0 \cdot 8113b$, $0 \cdot 649b$; and, when the momentum is $(\frac{4}{3}\pi)^{\frac{3}{2}} b^5 \times \cdot 644$, the axes are $3 \cdot 136b$, $0 \cdot 586b$, $0 \cdot 545b$.

It seems probable, then, that the Jacobian ellipsoid of mass $\frac{4}{3}\pi b^3$ becomes unstable, at least as soon as when the moment of momentum is somewhere about $(\frac{4}{3}\pi)^{\frac{3}{2}} b^5 \times \cdot 5$.

It may be worth mentioning that the greatest moment of momentum for which the ellipsoid (of mass $\frac{4}{3}\pi b^3$) is stable, when it is a figure of revolution, is $(\frac{4}{3}\pi)^{\frac{3}{2}} b^5 \times \cdot 3038$.

§ 11. *On the Conditions under which the two Masses may be close to one another.*

If at any point on the surface of either mass the sum of the tide-generating and centrifugal forces is greater than gravity, it is obvious that equilibrium cannot subsist. It is also clear that, if this condition is to be found anywhere, it will be at that point of the smaller mass which lies nearest to the larger mass. Hence, in order that the system may be a possible one, we must satisfy ourselves that at that point gravity of the body itself exceeds the sum of the tide-generating and centrifugal forces.

To determine the limitations of size and proximity of the smaller of the two masses to a high degree of approximation would be very laborious, and we shall, therefore, content ourselves with a rough investigation, to be explained below.

We shall now find approximations for the shapes of the two masses and for their potentials.

The radius-vector of either mass and the potential may be expanded in powers of a/c and A/c, and a term involving c^n in the denominator will be referred to as being of the nth order.

Now the term of the highest order which can be included without the introduction of great complication is the 7th, and we shall content ourselves with that term.

The expressions for the various parts of the potential have been developed above, but it may be observed that the terms involving the first order of harmonics may be

* 'Roy. Soc. Proc.,' vol. 41, 1887, p. 319.

omitted, since they are subsequently annulled by a proper choice of the angular velocity.

From (22–i.) we have

$$\frac{4\pi a^2}{3}\frac{a}{r} + \frac{4\pi A^3}{3c}\frac{3}{2}\left[h_2\left(\frac{a}{c}\right)^2\left(\frac{a}{r}\right)^3\frac{w_2}{r^2} + \tfrac{1}{2}h_3\left(\frac{a}{c}\right)^3\left(\frac{a}{r}\right)^4\frac{w_3}{r^3} + \ldots\right].$$

The last term in the development to the 7th order is that involving w_6. Then it is clear that we require h_2 correct to the 4th order, h_3 to the 3rd, and so on. But (25) shows us that the h's are equal to unity to the 4th order inclusive. Hence, in the above, all the h's may be treated as unity.

Again (22–ii.) when written in reference to the origin o affords other terms, in which all those included under $\Sigma\Sigma$ are of the 8th and higher orders, and negligible; and the rest (with omission of the first harmonic term) gives

$$\frac{4\pi A^3}{3c}\left[\left(\frac{a}{c}\right)^2\frac{w_2}{a^2} + \left(\frac{a}{c}\right)^3\frac{w_3}{a^3} + \ldots\right].$$

Thus this first part of the potential is, to the 7th order inclusive,

$$\frac{4\pi a^2}{3}\cdot\frac{a}{r} + \frac{4\pi A^3}{3c}\sum_{k=2}^{k=6}\left(\frac{a}{c}\right)^k\left\{\frac{3}{2(k-1)}\left(\frac{a}{r}\right)^{k+1} + \left(\frac{r}{a}\right)^k\right\}\frac{w_k}{r^k}.$$

Next, from the expression for Ω in § 3, we have a term in the potential due to rotation $+\frac{1}{3}\omega^2 r^2$. The remaining terms due to rotation will be taken up later.

From (71) we see that, to the 7th order inclusive,

$$\frac{3\omega^2}{4\pi} = \left(\frac{A}{c}\right)^3 + \left(\frac{a}{c}\right)^3.$$

Hence ω^2 and ϵ are of the 3rd order; and from (48) and (64) it follows that the factors by which the l's and m's are derived from the λ's and μ's are of the 5th order. And, since the λ's and μ's only differ from unity in terms of the 5th order, it follows that the l's and m's are of the 5th order. Then (41) and (56) show us that all the terms in l and m are negligible.

The first set of terms due to rotation and to the corresponding deformation are given in (39) and (43), and together contribute

$$\tfrac{1}{6}\omega^2 a^2\left[\tfrac{3}{2}\left(\frac{a}{r}\right)^3 + \left(\frac{r}{a}\right)^2\right]\frac{w_2}{r^2}.$$

The second set of terms due to rotation, and to the corresponding deformation, are given in (54) and (58), and together contribute

$$-\tfrac{1}{6}\omega^2 a^2\left[\tfrac{3}{2}\left(\frac{a}{r}\right)^3 + \left(\frac{r}{a}\right)^2\right]\tfrac{1}{2}\frac{\delta^2 w_4}{r^2}.$$

Hence, to the 7th order inclusive, we have

$$V = \frac{4\pi a^3}{3}\frac{a}{r} + \frac{4\pi A^3}{3c}\sum_{k=2}^{k=6}\left(\frac{a}{c}\right)^k\left\{\frac{3}{2(k-1)}\left(\frac{a}{r}\right)^{k+1} + \left(\frac{r}{a}\right)^k\right\}\frac{w_k}{r^k}$$
$$+ \tfrac{1}{3}\omega^2 a^2\left(\frac{r}{a}\right)^2 + \tfrac{1}{6}\omega^2 a^2\left[\tfrac{3}{2}\left(\frac{a}{r}\right)^3 + \left(\frac{r}{a}\right)^2\right]\left(\frac{w_2 - \frac{1}{2}\delta^2 w_4}{r^2}\right). \quad . \quad . \quad . \quad . \quad (78)$$

Now the expression (72) for the radius-vector of the mass a to the same order of approximation gives us

$$\frac{r}{a} = 1 + \frac{5\omega^2}{16\pi}\left(\frac{w_2 - \frac{1}{2}\delta^2 w_4}{r^3}\right) + \left(\frac{A}{c}\right)^3\sum_{k=2}^{k=6}\frac{2k+1}{2k-2}\left(\frac{a}{c}\right)^{k-2}\frac{w_k}{r^k},$$

and a similar expression for R/A.

To determine the inward force at the pole of the mass a, where it is nearest to the mass A, we must evaluate $-dV/dr$, and in the first term substitute the above expression for r, and in the remaining terms put $r = a$; also at this pole $w_2/r^2 = 1$, and $\delta^2 w_4 = 0$.

Then, differentiating (78),

$$-\frac{dV}{dr} = \frac{4\pi a}{3}\frac{a^3}{r^3} - \frac{4\pi a}{3}\left(\frac{A}{c}\right)^3\sum_{k=2}^{k=6}\left(\frac{a}{c}\right)^{k-2}\left\{-\frac{3(k+1)}{2(k-1)} + k\right\}$$
$$- \tfrac{2}{3}\omega^2 a - \tfrac{1}{6}\omega^2 a^3\left\{-\frac{3.3}{2} + 2\right\}. \quad . \quad . \quad . \quad . \quad (79)$$

But at the pole

$$\frac{4\pi a}{3}\frac{a^3}{r^3} = \frac{4\pi a}{3} - \tfrac{5}{6}\omega^2 a - \frac{4\pi a}{3}\left(\frac{A}{c}\right)^3\sum_{k=2}^{k=6}\frac{2k+1}{k-1}\left(\frac{a}{c}\right)^{k-2}.$$

Substituting this for the first term of (79), we have

$$-\frac{dV}{dr} = \frac{4\pi a}{3} - \tfrac{13}{12}\omega^2 a - \frac{4\pi a}{3}\left(\frac{A}{c}\right)^3\sum_{k=2}^{k=6}(k+\tfrac{1}{2})\left(\frac{a}{c}\right)^{k-2}.$$

But

$$\omega^2 a = \frac{4\pi a}{3}\left[\left(\frac{A}{c}\right)^3 + \left(\frac{a}{c}\right)^3\right];$$

hence

$$-\frac{dV}{dr} = \frac{4\pi a}{3}\left[1 - \tfrac{13}{12}\left[\left(\frac{A}{c}\right)^3 + \left(\frac{a}{c}\right)^3\right] - \left(\frac{A}{c}\right)^3\left\{\tfrac{5}{2} + \tfrac{7}{2}\frac{a}{c} + \tfrac{9}{2}\left(\frac{a}{c}\right)^2 + \tfrac{11}{2}\left(\frac{a}{c}\right)^3 + \tfrac{13}{2}\left(\frac{a}{c}\right)^4\right\}\right]$$
$$= \frac{4\pi a}{3}\left[1 - \frac{43A^3 + 13a^3}{12c^3} - \left(\frac{A}{c}\right)^3\left\{\tfrac{7}{2}\frac{a}{c} + \tfrac{9}{2}\left(\frac{a}{c}\right)^2 + \tfrac{11}{2}\left(\frac{a}{c}\right)^3 + \tfrac{13}{2}\left(\frac{a}{c}\right)^4\right\}\right].$$

Thus the criterion of the possibility of equilibrium is that

$$C = 1 - \frac{43A^3 + 13a^3}{12c^3} - \left(\frac{A}{c}\right)^3 \left\{ \frac{7}{2}\frac{a}{c} + \frac{9}{2}\left(\frac{a}{c}\right)^2 + \frac{11}{2}\left(\frac{a}{c}\right)^3 + \frac{13}{2}\left(\frac{a}{c}\right)^4 \right\} \quad . \quad . \quad (80)$$

should be positive.

But the radius-vectors of the poles are

$$\frac{r}{a} = 1 + \frac{5}{12}\left[\left(\frac{A}{c}\right)^3 + \left(\frac{a}{c}\right)^3\right] + \left(\frac{A}{c}\right)^3 \left[\frac{5}{2} + \frac{7}{4}\left(\frac{a}{c}\right) + \frac{9}{6}\left(\frac{a}{c}\right)^2 + \frac{11}{8}\left(\frac{a}{c}\right)^3 + \frac{13}{10}\left(\frac{a}{c}\right)^4\right],$$

and, similarly,

$$\frac{R}{A} = 1 + \frac{5}{12}\left[\left(\frac{A}{c}\right)^3 + \left(\frac{a}{c}\right)^3\right] + \left(\frac{a}{c}\right)^3 \left[\frac{5}{2} + \frac{7}{4}\left(\frac{A}{c}\right) + \frac{9}{6}\left(\frac{A}{c}\right)^2 + \frac{11}{8}\left(\frac{A}{c}\right)^3 + \frac{13}{10}\left(\frac{A}{c}\right)^4\right].$$

Therefore

$$r + R = a + A + \frac{5}{12c^3}[(7a + A)A^3 + (a + 7A)a^3] + \frac{7}{4c^4}A^2a^2(A + a)$$
$$+ \frac{3}{c^5}A^3a^3 + \frac{11}{8c^6}A^3a^3(a + A) + \frac{13}{10c^7}A^3a^3(a^2 + A^2).$$

Now the interval between the two masses is $c - (r + R)$; hence, if the two masses are just in contact,

$$c = a + A + \frac{5}{12c^3}[(7a + A)A^3 + (a + 7A)a^3] + \frac{7}{4c^4}A^2a^2(A + a)$$
$$+ \frac{3}{c^5}A^3a^3 + \frac{11}{8c^6}A^3a^3(a + A) + \frac{13}{10c^7}A^3a^3(a^2 + A^2). \quad . \quad (81)$$

In order, then, to test whether equilibrium is still possible when the two masses are just in contact, it is necessary to determine c from (81); and then, substituting in (80), find whether C is positive or not.

The solution of an equation

$$c = \alpha + \frac{\beta}{c^3} + \frac{\gamma}{c^4} + \frac{\delta}{c^5} + \frac{\epsilon}{c^6} + \frac{\zeta}{c^7},$$

and the determination of

$$C = 1 - \frac{\mathrm{B}}{c^3} - \frac{\Gamma}{c^4} - \frac{\Delta}{c^5} - \frac{\mathrm{E}}{c^6} - \frac{\mathrm{Z}}{c^7}$$

can only be performed by trial and error.

Now suppose that the solution is $c_0 + \delta c$, where δc is small; and that

$$c_1 = \alpha + \frac{\beta}{c_0^3} + \frac{\delta}{c_0^4} + \ldots, \qquad C_1 = 1 - \frac{\mathrm{B}}{c_0^3} - \frac{\Gamma}{c_0^4} - \ldots$$

Then it is obvious that

$$\frac{\delta c}{c_0} = \frac{c_1 - c_0}{c_0 + \frac{3\beta}{c_0^{\,3}} + \frac{4\gamma}{c_0^{\,4}} + \ldots}$$

and

$$\delta C = \left(\frac{3B}{c_0^{\,3}} + \frac{4\Gamma}{c_0^{\,4}} + \ldots\right)\frac{\delta c}{c_0}, \qquad \text{and} \qquad C = C_1 + \delta C$$

It is not hard to find an approximate solution c_0 by trial and error, and the correct results may then be found thus.

Consider the case where the two bodies are equal to one another, and put $a = A = 1$. The equations then become

$$c = 2 + \frac{20}{3c^3} + \frac{7}{2c^4} + \frac{3}{c^5} + \frac{11}{4c^6} + \frac{13}{5c^7},$$

$$C = 1 - \frac{14}{3c^3} - \frac{7}{2c^4} - \frac{9}{2c^5} - \frac{11}{2c^6} - \frac{13}{2c^7}.$$

By trial and error we find $c = 2\cdot535$, $C = +\cdot557$.

From this we conclude that equilibrium still subsists when the two masses are in contact.

When $a = A = 1, c = 2\cdot535$, we have $\gamma = (a/c)^2 = \frac{1}{6\cdot43}$, and $\beta = \gamma/(1-\gamma) = \frac{1}{5\cdot43}$. Our figures showed that when $\beta = \frac{1}{6}$ the two masses were nearly in contact, and when $\beta = \frac{1}{5}$ they crossed.

This result is, therefore, in accordance with the figures.

Next pass to the case of an infinitesimal satellite, and suppose a infinitely small compared with A and c, and that $A = 1$. The equations are

$$c = 1 + \frac{5}{12c^3},$$

$$C = 1 - \frac{43}{12c^3}.$$

The solution of the first equation is $c = 1\cdot226$, and this value of c makes $C = -\cdot94$. Hence we conclude that an infinitesimal fluid satellite cannot revolve with its surface in contact with its planet.

C vanishes when $c = (\frac{43}{12})^{\frac{1}{3}} = 1\cdot89$. Hence it appears that the nearest approach of the infinitesimal satellite to the planet is $1\cdot89$ mean radii of the planet. The nature of the approximation adopted is, however, such that in reality the satellite must lie further from the planet than this, perhaps at two radii distance.* The satellite and planet of which we here speak are, of course, supposed to revolve as parts of a rigid

* [See Roche, 'Montpellier, Acad. Sci. Mém.,' vol. 1, 1847–1850, p. 243 (added Oct. 5, 1887).]

body. Now, if for equal masses equilibrium still subsists when the two masses are in contact, whilst for infinitesimal mass of one equilibrium is impossible with the masses in contact, it follows that for some ratio of masses equilibrium can just subsist when they are in contact.

The question, therefore, remains to determine this limiting ratio of masses.

I find, then, that when $a = 1$, $A = 3{\cdot}4$, we have

$$c = 4{\cdot}4 + [2{\cdot}25684]c^{-3} + [1{\cdot}94945]c^{-4} + [2{\cdot}07156]c^{-5} + [2{\cdot}37619]c^{-6} + [2{\cdot}80737]c^{-7},$$
$$C = 1 - [2{\cdot}15205]c^{-3} - [2{\cdot}13850]c^{-4} - [2{\cdot}24765]c^{-5} - [2{\cdot}33480]c^{-6} - [2{\cdot}40735]c^{-7},$$

the numbers in [ ] being the logarithms of the coefficients.

The solution of this is $c = 5{\cdot}57$, which makes $C = -\ {\cdot}006$.

Again, when $a = 1$, $A = 3{\cdot}3$, we have

$$c = 4{\cdot}3 + [2{\cdot}21556]c^{-3} + [1{\cdot}91353]c^{-4} + [2{\cdot}03266]c^{-5} + [2{\cdot}32731]c^{-6} + [2{\cdot}74467]c^{-7},$$
$$C = 1 - [2{\cdot}11347]c^{-3} - [2{\cdot}09961]c^{-4} - [2{\cdot}20876]c^{-5} - [2{\cdot}29591]c^{-6} - [2{\cdot}36846]c^{-7},$$

the solution of which is $c = 5{\cdot}45$, which makes $C = +\ {\cdot}010$. Since $(3{\cdot}4)^3 = 39{\cdot}3$, and $(3{\cdot}3)^3 = 35{\cdot}9$, it follows that the ratio of the masses in the first case is $1 : 39{\cdot}3$, and in the second $1 : 35{\cdot}9$.

From this it appears that when the ratio of the masses is about 1 to 38 equilibrium is still just possible when the two masses touch.

It must be borne in mind, however, that the nature of the approximations adopted in this investigation is such that the results in this limiting case are only given very roughly, and it is certain that actually the limiting size of the smaller of the two masses must be greater than as thus computed.

We can only conclude that the limiting case occurs when the ratio of the masses is about 1 to 30, or the ratio of the radii about 1 to 3.

There is one other case which it is interesting to consider, namely, to find the limiting proximity of the Moon to the Earth, both bodies being treated as homogeneous fluids of the same density, revolving as a rigid body.

The case of Moon and Earth is well represented by $a = 1$, $A = 4{\cdot}333$; for this gives 1 to $81{\cdot}35$ as the ratio of the masses. With these values I find

$$C = 1 - [2{\cdot}46626]c^{-3} - [2{\cdot}45443]c^{-4} - [2{\cdot}56358]c^{-5} - [2{\cdot}65073]c^{-6} - [2{\cdot}72328]c^{-7},$$
and
$$r + R = 5{\cdot}333 + [2{\cdot}59898]c^{-3} + [2{\cdot}24358]c^{-4} + [2{\cdot}38748]c^{-5} + [2{\cdot}77563]c^{-6} + [3{\cdot}32042]c^{-7}.$$

Now $c = 7{\cdot}0$ will be found to make C vanish, and, with this value of c, $c - (r + R) = {\cdot}414$.

If A be 4000 miles, $c = 6500$ miles, and $c - (r + R) = 380$ miles.

Thus, as far as this investigation goes, it appears that when the fluid Moon is on the point of breaking up from stress of tidal and centrifugal forces the distance between the centres of Moon and Earth is 6500 miles, and the shortest distance between the two surfaces is 380 miles.

This result must, however, from the nature of the approximation, be an under-estimate of the distances.

The whole of the present section has been suggested by a pamphlet by Mr. JAMES NOLAN * in which he criticises some of my previous papers. I have commented elsewhere on his criticisms.†

§ 12. *On the Case where the two Masses are unequal.*

The results of the previous section point to a very remarkable limitation to the possibility of approach of two masses of unequal size. It has, therefore, seemed worth while to consider this case numerically, and a case is therefore chosen which shall approach near to that which we know is the limit of possibility. I choose, therefore, $a = 1$, $A = 3$, which makes the ratio of the masses 1 to 27, and $c = 5{\cdot}3$, which brings the protuberances into close proximity.

The numerical details are omitted, but figs. 5 and 6 (Plate 23) give the results, the numerical values of the radius-vectors being, as before, entered on the figure.

The elongation of the smaller mass is so extreme that it is obvious that, rigorously speaking, the spherical harmonic approximation must be considered to break down. Nevertheless, I conceive that these curious figures may be held to indicate the general nature of the true result.

It is remarkable that the smaller mass exhibits a marked furrowing round the middle. This seems to indicate that such a system tends to break up by the separation of the smaller mass into two parts.

§ 13. *Summary.*

The intention of this paper is, first, to investigate the forms which two masses of fluid assume when they revolve in close proximity about one another, without relative motion of their parts; and, secondly, to obtain a representation of the single form of equilibrium which must exist when the two masses approach so near to one another as just to coalesce into a single mass.

When the two masses are far apart the solution of the problem is simply that of the equilibrium theory of the tides. Each mass may, as far as its action on the other is

* 'DARWIN'S Theory of the Genesis of the Moon.' ROBERTSON, Melbourne, Sydney, Adelaide, and Brisbane, 1885.

† 'Nature,' February 18 and July 29, 1886.

concerned, be treated as spherical, and the tide-generating potential is given with sufficient accuracy by a single term of the second order of harmonics. As the masses are brought nearer to one another, this approximation ceases to be sufficient, terms of higher orders of harmonics become necessary to represent the potential adequately, and the departure from sphericity of each mass begins to exercise a sensible deforming influence on the other.

When the departure from sphericity of one body produces a sensible deformation in the other, that deformation in its turn reacts on the first, and thus the actual figure assumed by either mass may be regarded as a deformation due to the primitive influence of the other mass, on which is superposed the sum of an infinite series of reflected deformations.

But each mass is deformed, not only by the tidal action of the other, but also by its own rotation about an axis perpendicular to its orbit. The departure from sphericity of either body due to rotation also exercises an influence on the other, and thus there arises another infinite series of reflected deformations. It is shown in this paper how the summations of these two kinds of reflections are to be made by means of the solution of three sets of linear equations for the determination of three sets of coefficients.

The first set of coefficients are augmenting factors, by which the tides of each order of harmonics are to be raised above the value which they would have if the perturbing mass were spherical. It appears that, the higher the order of harmonics, the more do these factors exceed unity.

The second set of coefficients correspond to one part of the rotational effects. They appertain to terms of exactly the same form as the tidal terms, and in the final result the terms to which they apply become fused with the tidal terms. These terms are the zonal harmonics of the several orders with respect to the axis joining the centres of the two masses.

The third set of coefficients correspond to the remainder of the rotational effect, and they appertain to a different kind of deformation. These deformations are represented by sectorial harmonics involving $\cos 2\phi$, where ϕ is azimuth measured from the plane passing through the axis of rotation of the system and the centres of the two masses. That term of this set which is of the second order of harmonics, and which represents the ellipticity of either mass augmented by mutual influence, is the only term which is considerable, even when the two masses are very close together; but the existence of the other harmonic deformations of this class is interesting. We may say, then, that all the tides of either mass are augmented above the values which they would have if the other mass were spherical; that the ellipticity corresponding to rotation is augmented; and that the deformation due to rotation is no longer exactly elliptic-spheroidal.

The angular velocity of the system is found by the consideration that the repulsion due to centrifugal force between the two masses shall exactly balance the resultant attraction between them. If the masses were spherical, the square of the orbital angular velocity, multiplied by the cube of the distance between the centres, would be equal to the sum of the masses. When the masses are deformed, however, this law is no longer true, and the angular velocity has to be augmented by a factor a little greater than unity, which depends on the amounts of the deformations.

The theory here sketched is applied above numerically to several cases, and the results will be found in the preceding paragraphs. We shall first consider the cases where the two masses are equal to one another.

In the first example ($\beta = \frac{1}{7}$) solved numerically, the distance between the centres of the two masses is 2·83 times the mean radius of either of them. The two bodies are found to be elongated until they approach near to one another; but, as the character of the distortion is better illustrated in a subsequent case, the result is not given graphically. All the data, however, are found which will enable the reader to draw the figure if he should wish to do so.

In the next example ($\beta = \frac{1}{6}$), with the masses still equal, the distance between the centres is reduced to 2·646 of the mean radius of either. The result of the solution is illustrated by two figures. In fig. 4, Plate 22, the section of the masses by a plane perpendicular to the axis of rotation is shown, and in fig. 5, Plate 23, we have the section by a plane passing through the axis and the centres of the two masses. On both figures are inscribed the values of the radii for each 15° of latitude in terms of the mean radius as unity, and the mean sphere, from which the distortion is computed, is marked by a short line on each radius. The elongation of the masses is, of course, considerably greater in the section through the axis than in the other section. Each mass is shaped somewhat like an egg, and the small ends face one another and come very nearly into contact.

In the headings to the figures, amongst other numerical data, are given the square of the angular velocity and the angular momentum of the system. The density of the fluid being unity, the angular velocity ω is given by the value of $\omega^2/4\pi$; this is the function of angular velocity which is usually given when reference is made to figures of equilibrium of rotating fluid, such as the revolutional or Jacobian ellipsoids of equilibrium. The moment of momentum of the system is given by reference to the angular velocity of a sphere, of the same mass as the sum of our two masses, rotating so as to have the same momentum. If, in fact, b be such a length that a sphere of fluid of that radius has the same mass as our system (so that $b^3 = a^3 + A^3$), then the moment of momentum is given by a number μ in the expression $(\frac{4}{3}\pi)^{\frac{3}{2}} b^5 \times \mu$. By this notation the angular velocity and moment of momentum are made comparable with the results given in a previous paper* on the Jacobian ellipsoid of equilibrium. From that paper the following table of the axes, angular velocity, and moment of momentum of several solutions of JACOBI's problem is extracted.

* 'Roy. Soc. Proc.', vol. 41, 1887, p. 319.

JACOBI'S ELLIPSOIDS.

	Axes.			Ang. vel.	Mom.
	Greatest $\frac{a}{b}$	Mean $\frac{b}{b}$	Least $\frac{c}{b}$	$\frac{\omega^2}{4\pi}$	μ
1	1·1972	1·1972	·6977	·09356	·30375
2	1·279	1·123	·696	·093	·306
3	1·383	1·045	·692	·0906	·3134
4	1·601	·924	·677	·0830	·3407
5	1·899	·811	·649	·0705	·3920
6	2·346	·702	·607	·0536	·4809
7	3·136	·586	·545	·0334	·644
8	5·04	·45	·44	·013	1·016
9	∞	·00	·00	·000	∞

In figs. 4 and 5, Plates 22, 23, $\omega^2/4\pi$ is ·038, and the momentum μ is ·472. On comparison with the Table of JACOBI's ellipsoids, we see that this corresponds with a considerably slower rotation than the 6th solution, and nearly the same moment of momentum.

In the next case the two masses are still closer ($\beta = \frac{1}{5}$), the distance between the centres being only 2·449 times either mean radius. The result is illustrated in figs. 2 and 3; the explanation of figs. 4 and 5 serves, *mutatis mutandis*, for these two figures also.

This case is interesting because the masses have approached so near to one another that they partially overlap. Two portions of matter cannot, of course, occupy the same space, and the continuity of figures of equilibrium leads us to believe that the reality must consist of a single mass of fluid. In figs. 2 and 3 conjectural dotted lines are drawn to show how it is probable that the overlapping of the two masses is replaced by a neck of fluid joining them. The figures as thus amended serve to give a good representation of the single dumb-bell shaped figure of equilibrium.

The angular velocity is here given by $\omega^2/4\pi =$ ·049, and the moment of momentum by ·482. In the sixth entry of the Table of Solutions of JACOBI's problem we find $\omega^2/4\pi =$ ·0536, and the moment of momentum $\mu =$ ·481. This ellipsoid has, then, the same moment of momentum, and only about 4 per cent. more angular velocity, than our dumb-bell. It has seemed, therefore, worth while to mark (in chain-dot) on figs. 2 and 3 the outline of this Jacobian ellipsoid of the same mass as the dumb-bell. The actual vertex of the ellipsoid just falls outside the limits to which it was possible to extend the figure.

In the paper above referred to it is shown how the energy of the Jacobian ellipsoid is to be computed. If we denote the kinetic energy by $(\frac{4}{3}\pi)^2 b^5 \times \epsilon$, and the intrinsic energy by $(\frac{4}{3}\pi)^2 b^5 \times (i - 1)$,* then it appears that in the case of the ellipsoid drawn in these figures $\epsilon =$ ·0964, $i =$ ·4808, and the total energy $E = \epsilon + i =$ ·5772.

* The intrinsic energy being negative, it is more convenient to tabulate i a positive quantity.

Now in the case of our dumb-bell figure it appears, from calculations referred to in the Appendix, that $\epsilon = \cdot 0925$, $i = \cdot 4873$, and $E = \cdot 5798$. Hence in the dumb-bell figure the kinetic energy is less, but the intrinsic energy is so much greater that the total energy is about a half per cent. greater. These numbers are, of course, computed from the approximate formulæ, and must not be taken as rigorously correct for the dumb-bell figure of equilibrium.

With reference to a figure of transition from the Jacobian ellipsoid, Sir WILLIAM THOMSON has remarked :—*

"We have a most interesting gap between the unstable Jacobian ellipsoid, when too slender for stability, and the case of smallest moment of momentum consistent with stability in two equal detached portions. The consideration of how to fill up this gap with intermediate figures is a most attractive question, towards answering which we at present offer no contribution." †

Figs. 2 and 3 are intended to form such a contribution, but it is certain that the matter is far from being probed to the bottom.

M. POINCARÉ has made an admirable investigation of the forms of equilibrium of a single rotating mass of fluid, and has especially considered the stability of JACOBI's ellipsoid. ‡ He has shown, by a difficult analytical process, that when the ellipsoid is moderately elongated (he has not arrived at a numerical result) instability sets in by a furrowing of the ellipsoid along a line which lies in a plane perpendicular to the longest axis. It is, however, extremely remarkable that the furrow is not symmetrical with respect to the two ends, and thus there appears to be a tendency to form a dumb-bell with unequal bulbs.

If M. POINCARÉ's result shall appear to be not only true, but to contain the whole truth concerning the mode in which instability of the ellipsoid supervenes, then there must be some other transitional form between the unsymmetrically furrowed Jacobian

* THOMSON and TAIT's 'Natural Philosophy,' 1883, § 778″ (*i*).

† In 778″ (*g*) he remarks that "a deviation from the ellipsoidal figure in the way of thinning it in the middle and thickening it towards the end would, with the same moment of momentum, give less energy." I conceive that the energy referred to throughout this paragraph is kinetic only, and we have seen that the kinetic energy is less for the dumb-bell than for the ellipsoid.

[If we write U for a quantity proportional to the excess of kinetic above intrinsic energy, so that $U = \epsilon + (1 - i)$, then figures of equilibrium are to be determined by making U stationary for variations of the parameters involved in it. This course is actually pursued in the Appendix below, the function (viii.) being, in fact, this U; and the variations of it, being made stationary, afford a controlling solution of the problem of this paper. The similar method may easily be applied to the case of JACOBI's ellipsoids. From this point of view the interesting function to tabulate is $\epsilon + (1 - i)$, and we observe that in the case of the Jacobian ellipsoid referred to on the last page it is ·6052, and for the dumb-bell it is ·6156. Is not the energy referred to by Sir W. THOMSON this function U? (*Addition to foot-note, dated October* 10, 1887.)]

‡ 'Acta Mathematica,' 7, 3 and 4, 1885.

ellipsoid and the dumb-bell; except, perhaps, in the case where the two bulbs pass on to two masses of a definite ratio.

M. POINCARÉ's work seemed so important that this paper was kept back for a year, whilst I endeavoured to apply the principles, which he has pointed out, to the discussion of the stability of the two masses. The attempt, which is given in the Appendix, is apparently abortive, on account of the imperfections of spherical harmonic analysis when applied to bodies which depart considerably from the spherical shape.

We must, therefore, leave this complex question in abeyance, and merely point to the Appendix as an example of the method which must almost certainly be pursued if this problem is to yield its answer to analysis.

Allusion has just been made to the imperfection of spherical harmonic analysis, and this brings us naturally to face the question whether that analysis may not have been pushed altogether too far in the computation of the figures of equilibrium under discussion. This question is considered in § 9, and a rough criterion of the limits of applicability of this analysis is there found. From this it appears that even in the cases of figs. 2 and 3 the result must present a fair approximation to correctness. The criterion, indeed, appears to be such as necessarily to give too unfavourable a view of the correctness of the result.

The rigorous method of discussing the stability of the system having failed, certain considerations are adduced in § 11 which bear on the conditions under which there is a form of equilibrium consisting of two fluid masses in close promixity. It appears that there cannot be such a form with the two masses just in contact, unless the smaller of the two masses exceeds in mass about one-thirtieth of the larger.

If we take into consideration the fact that the criterion of the applicability of harmonic analysis is too severe, it appears to be worth while to find to what results the analysis leads when two masses, one 27 times as great as the other, are brought close together. The numerical work of the calculation is omitted, since the numbers can only represent the true conclusion very roughly; but the result is illustrated graphically in figs. 6 and 7, Plate 23. These figures can only serve to give a general idea of the truth, but the form into which the smaller mass is thrown is so remarkable as to be worthy of attention. The deep furrow round the smaller mass, lying in a plane parallel to the axis of rotation, cannot be due merely to the imperfection of the solution; and it appears to point to the conclusion that there is a tendency for the smaller body to separate into two, just as we have seen the Jacobian ellipsoid become dumb-bell shaped and separate into two parts.

In this paper, indeed, we have sought to trace the process in the opposite direction, and to observe the coalescence of two masses into one. The investigation is complementary to, but far less perfect than, that of M. POINCARÉ, who describes the series of changes which he has been tracing in the following words:—

"Considérons une masse fluide homogène animée originairement d'un mouvement de rotation; imaginons que cette masse se contracte en se refroidissant lentement, mais de façon à rester toujours homogène. Supposons que le refroidissement soit assez lent et le frottement intérieur du fluide assez fort pour que le mouvement de rotation reste le même dans les diverses portions du fluide. Dans ces conditions le fluide tendra toujours à prendre une figure d'équilibre séculairement stable. Le moment de la quantité de mouvement restera d'ailleurs constant.

"Au début, la densité étant très faible, la figure de la masse est un ellipsoïde de révolution très peu différent d'une sphère. Le refroidissement aura d'abord pour effet d'augmenter l'aplatissement de l'ellipsoïde, qui restera cependant de révolution. Quand l'aplatissement sera devenu à peu près égal à $\frac{2}{5}$, l'ellipsoïde cessera d'être de révolution et deviendra un ellipsoïde de JACOBI. Le refroidissement continuant, la masse cessera d'être ellipsoïdale; elle deviendra dissymétrique par rapport au plan des yz, et elle affectera la forme représentée dans la figure, p. 347.*

"Comme nous l'avons fait observer à propos de cette figure, l'ellipsoïde semble se creuser légèrement dans sa partie moyenne, mais plus près de l'un des deux sommets du grand axe; la plus grande partie de la matière tend à se rapprocher de la forme sphérique, pendant que la plus petite partie sort de l'ellipsoïde par un des sommets du grand axe, comme si elle cherchait à se détacher de la masse principale.

"Il est difficile d'annoncer avec certitude ce qui arrivera ensuite si le refroidissement continue, mais il est permis de supposer que la masse ira en se creusant de plus en plus, puis en s'étranglant dans la partie moyenne, et finira par se partager en deux corps isolés.

"On pourrait être tenté de chercher dans ces considérations une confirmation ou une réfutation de l'hypothèse de LAPLACE, mais on ne doit pas oublier que les conditions sont ici très différentes, car notre masse est homogène, tandis que la nébuleuse de LAPLACE devait être très fortement condensée vers le centre." †

It was in the hope that the investigation might throw some light on the nebular hypothesis of LAPLACE and KANT that I first undertook the work. It must be admitted, however, that we do not obtain much help from the results. It is justly remarked by M. POINCARÉ that the conditions for the separation of a satellite from a nebula differ from those of his problem in the great concentration of density in the central body. But both his investigation and the considerations adduced here seem to show that, when a portion of the central body becomes detached through increasing angular velocity, the portion should bear a far larger ratio to the remainder than is observed in the satellites of our system as compared with their planets; and it is hardly probable that the heterogeneity of the central body can make so great a difference in the result as would be necessary if we are to make an application of these ideas. It appears then at present necessary to suppose that after the birth of a satellite, if it takes place at all in this way, a series of changes occur which are quite unknown.

* The furrowed ellipsoid of JACOBI.

† POINCARÉ, 'Acta Mathemat.,' 7, 1885, p. 379.

APPENDIX.

On the Energy and Stability of the System.

M. POINCARÉ has shown in his admirable memoir, referred to in the Summary, how the dynamical stability of a rotating fluid system in relative equilibrium depends on the energy. Certain factors in the expression for the energy, which he calls coefficients of stability, are there proved to afford the required criterion.

It will now be shown how in this case these coefficients of stability are determinable, at least as far as spherical harmonic analysis permits. The results will also cast an interesting light on the methods by which the equations to the two masses have been obtained.

The task before us is to determine the "exhaustion of potential energy" of the two masses in presence of one another as due to the deformation of each from the spherical figure by yielding to gravitation and to centrifugal force.

The work will be rendered simpler by the introduction of a new notation. Let us write, then, as the equations to two shapes, which are not necessarily together a figure of equilibrium :—

$$\left.\begin{aligned}\frac{r}{a} &= 1 + \sum_{k=2}^{k=\infty} \frac{2k+1}{2k-2}\left(\frac{A}{c}\right)^3\left(\frac{a}{c}\right)^{k-2}\left\{n_k\frac{w_k}{r^k} - p_k\frac{\delta^2 w_{k+2}}{r^k}\right\}\\ \frac{R}{A} &= 1 + \sum_{k=2}^{k=\infty} \frac{2k+1}{2k-2}\left(\frac{a}{c}\right)^3\left(\frac{A}{c}\right)^{k-2}\left\{N_k\frac{W_k}{R^k} - P_k\frac{\delta^2 W_{k+2}}{R^k}\right\}\end{aligned}\right\} \quad . \quad . \quad . \quad . \quad . \quad . \quad \text{(i.)}$$

It will be observed that these equations have the same form as (72), but that the constants introduced are different from the h, l, m, e, which were determined, so that the figures might be figures of equilibrium. At present we do not assume that (i.) do represent figures of equilibrium.

The energy lost may be divided into several parts :—

e_1, the energy lost by the mass a yielding from its spherical figure to the gravitation of the mean sphere a.

e_2, the exhaustion of mutual energy of that layer of matter on the mass a which constitutes its departure from sphericity.

e_3, the loss of energy due to the deformation of the mass a in presence of the mean sphere A.

E_1, E_2, E_3, the similar quantities for the mass A.

$(Ee)_4$, the loss of mutual energy of the two layers in presence of one another.

e_5, the loss of energy due to the deformation of the mass a in the presence of centrifugal force due to rotation ω.

E_5, the similar loss for A.

1st. e_1 is equal and opposite to the work required to raise each element of the layer on a through half its own height against the gravity due to the mean sphere a. This gravity is $\frac{4}{3}\pi a$. Co-latitude and longitude being denoted by θ, ϕ, let $d\varpi = \sin\theta\, d\theta\, d\phi$, an element of solid angle. In effecting the integrations, the properties of spherical harmonic functions are used without comment, viz. :—

$$\iint\left(\frac{w_k}{r^k}\right)^2 d\varpi = \frac{4\pi}{2k+1}, \qquad \iint\left(\frac{\delta^2 w_{k+2}}{r^k}\right)^2 d\varpi = \frac{4\pi}{2k+1}\,\frac{k+2!}{2\,.\,k-2!}, \qquad \iint w_k\,\delta^2 w_{k+2}\, d\varpi = 0,$$

$$\iint w_i w_k\, d\varpi = 0, \qquad \iint w_i\,\delta^2 w_{k+2}\, d\varpi = 0.$$

Then, taking only a typical term of the first of (i.),

$$\begin{aligned} e_1 &= -\iint \tfrac{4}{3}\pi a^6 \,.\, \tfrac{1}{2} \,.\, \left[\frac{2k+1}{2k-2}\left(\frac{A}{c}\right)^3\left(\frac{a}{c}\right)^{k-2}\left\{n_k\frac{w_k}{r^k} - p_k\frac{\delta^2 w_{k+2}}{r^k}\right\}\right]^2 d\varpi \\ &= -\left(\frac{4\pi}{3}\right)^2 \tfrac{1}{2}a^5\,\frac{2k+1}{(2k-2)^2}\,3\left(\frac{A}{c}\right)^6\left(\frac{a}{c}\right)^{2k-4}\left[n_k^2 + \frac{k+2!}{2\,.\,k-2!}\,p_k^2\right],\end{aligned}$$

whence, with all the terms, and remembering that $(a/c)^2 = \gamma$, $(A/c)^2 = \Gamma$,

$$e_1 = -\tfrac{1}{2}\left[\left(\frac{4\pi}{3}\right)^2 \frac{A^3a^3}{c}\right] \tfrac{3}{2} \left(\frac{A}{c}\right)^3 \sum_{k=2}^{k=\infty} \frac{2k+1}{2k-2}\,\frac{\gamma^{k-1}}{k-1}\left[n_k^2 + \frac{k+2!}{2\,.\,k-2!}\,p_k^2\right]. \quad . \quad . \quad . \quad \text{(ii.)}$$

The formula for E_1 may be written down by symmetry.

2nd. e_2, the exhaustion of mutual energy of the layer on itself, is half the potential of the layer at any element, multiplied by the mass of the element, and integrated over the whole sphere.

The potential of the layer is

$$\frac{4\pi A^3}{3c} \sum_{k=0}^{k=\infty} \frac{3}{2k-2} \left(\frac{a}{c}\right)^k \left(\frac{a}{r}\right)^{k+1} \left(n_k \frac{w_k}{r^k} - p_k \frac{\partial^2 w_{k+2}}{r^k}\right).$$

Then, at an element of the layer $r = a$, and taking a typical term only, we have

$$e_2 = \tfrac{1}{2} \cdot \frac{4\pi A^3}{3c} \cdot \frac{3(2k+1)}{(2k-2)^2} \left(\frac{a}{c}\right)^{2k-2} \left(\frac{A}{c}\right)^3 a^3 \iint \left[n_k \left(\frac{w_k}{r^k}\right)^2 + p_k \left(\frac{\partial^2 w_{k+2}}{r^k}\right)^2\right] d\varpi,$$

whence

$$e_2 = \tfrac{1}{2}\left[\left(\frac{4\pi}{3}\right)^2 \frac{A^3a^3}{c}\right] \tfrac{3}{2} \left(\frac{A}{c}\right)^3 \sum_{k=2}^{k=\infty} \frac{3}{2k-2}\,\frac{\gamma^{k-1}}{k-1}\left[n_k^2 + \frac{k+2!}{2\,.\,k-2!}\,p_k^2\right]. \quad . \quad . \quad . \quad \text{(iii.)}$$

The formula for E_2 may be written down by symmetry.

The addition of e_1 to e_2, and of E_1 to E_2, simplifies these expressions by cutting out the factor immediately following the Σ in either, and replacing it by unity.

3rd. e_3 is the loss of energy due to raising the layer on a in presence of the mean sphere A. We multiply the potential of the sphere A by the mass of the element on a, and integrate throughout angular space.

The potential of the sphere A, when transferred to the origin o, is

$$\frac{4\pi A^3}{3c} \sum_{k=0}^{k=\infty} \left(\frac{a}{c}\right)^k \left(\frac{r}{a}\right)^k \frac{w_k}{r^k}.$$

Then, at an element of the layer $r = a$ and taking a typical term,

$$e_3 = \frac{4\pi A^3}{3c}\,\frac{2k+1}{2k-2} \left(\frac{A}{c}\right)^3 \left(\frac{a}{c}\right)^{2k-2} a^3 \iint n_k \left(\frac{w_k}{r^k}\right)^2 d\varpi,$$

whence

$$e_3 = \left(\frac{4\pi}{3}\right)^2 \frac{A^3a^3}{c}\,\tfrac{3}{2} \left(\frac{A}{c}\right)^3 \sum_{k=2}^{k=\infty} \frac{\gamma^{k-1}}{k-1}\,n_k. \quad . \quad . \quad . \quad . \quad . \quad . \quad . \quad . \quad . \quad . \quad \text{(iv.)}$$

The expression for E_3 may be written down by symmetry. On collecting results from (ii.), (iii.), and (iv.), we have

$$e_1 + e_2 + e_3 = \left(\frac{4\pi}{3}\right)^2 \frac{A^3a^3}{c} \cdot \tfrac{3}{2} \left(\frac{A}{c}\right)^3 \sum_{k=2}^{k=\infty} \frac{\gamma^{k-1}}{k-1} \left\{n_k - \tfrac{1}{2}n_k^2 - \frac{k+2!}{4\,.\,k-2!}\,p_k^2\right\}, \quad . \quad . \quad . \quad \text{(v.)}$$

and a similar expression for $E_1 + E_2 + E_3$.

4th. $(Ee)_4$ is the loss of energy of one layer in the presence of the other. We take the potential of the layer on A, multiply it by the mass of an element on a, and integrate.

The potential of the layer on A when transferred to the origin o, as in (22-ii.), is

$$\frac{4\pi A^3}{3c}\,\tfrac{3}{2} \left(\frac{a}{c}\right)^3 \left\{\sum_{k=2}^{k=\infty} \sum_{i=2}^{i=\infty} \left[\frac{k+i!}{k!\,i!}\,\frac{\Gamma^{i-1}}{i-1} N_i \left(\frac{a}{c}\right)^k \left(\frac{r}{a}\right)^k \frac{w_k}{r^k} - \frac{k+i!}{i-2!\,k+2!}\,\frac{\Gamma^{i-1}}{i-1} P_i \left(\frac{a}{c}\right)^k \left(\frac{r}{a}\right)^k \frac{\partial^2 w_{k+2}}{r^k}\right]\right\}.$$

Introducing this into the integral, only taking a typical term, and neglecting those terms in the integral which must vanish, we get

$$(Ee)_4 = \frac{4\pi A^3}{3c}\,\frac{3}{2}\left(\frac{a}{c}\right)^3 \frac{2k+1}{2k-2}\left(\frac{A}{c}\right)^3\left(\frac{a}{c}\right)^{2k-2} a^3\,\frac{\Gamma^{i-1}}{i-1}\left\{\iint \frac{k+i!}{k!\,i!}\,N_i n_k\left(\frac{w_k}{r^k}\right)^2 d\varpi \right.$$
$$\left. + \iint \frac{k+i!}{i-2!\,k+2!}\,P_i p_k\left(\frac{\delta^2 w_{k+2}}{r^k}\right)^2 d\varpi\right\}.$$

Then, effecting the integrations, and putting in the $\Sigma\Sigma$, we get

$(Ee)_4 =$

$$\left(\frac{4\pi}{3}\right)^2 \frac{A^3 a^3}{c}\cdot\left(\frac{3}{2}\right)^2\left(\frac{a}{c}\right)^3\left(\frac{A}{c}\right)^3 \sum_{k=2}^{k=\infty}\sum_{i=2}^{i=\infty} \frac{\Gamma^{i-1}}{i-1}\,\frac{\gamma^{k-1}}{k-1}\left\{\frac{k+i!}{k!\,i!}N_i n_k + \frac{k+2!}{2.k-2!}\,\frac{k+i!}{i-2!\,k+2!}P_i p_k\right\}. \quad \text{(vi.)}$$

This involves the two figures symmetrically.

5th. e_5 is the loss of energy in the yielding of the figure a to centrifugal force. To find it, we multiply the potential due to rotation by the mass of each element of the layer a, and integrate.

By (35) and (52) we know that the rotation potential is

$$\tfrac{1}{6}\omega^2 a^2\left(\frac{r}{a}\right)^2\left\{\frac{w_2}{r^2} - \frac{1}{2}\,\frac{\delta^2 w_4}{r^2}\right\}.$$

As this only involves harmonics of the second order, we may neglect in the layer a all terms except those of the second order. Thus we get

$$e_5 = \tfrac{1}{6}\omega^2 a^2\,\tfrac{3}{2}\left(\frac{A}{c}\right)^3 a^3 \iint \left\{n_2\left(\frac{w_2}{r^2}\right)^2 + \tfrac{1}{2}p_2\left(\frac{\delta^2 w_4}{r^2}\right)^2\right\} d\varpi$$
$$= 4\pi\,.\,\tfrac{1}{12}\omega^2 a^5\left(\frac{A}{c}\right)^3\left\{n_2 + \frac{4!}{2.2.0!}p_2\right\}$$
$$= \left(\frac{4\pi}{3}\right)^2 \frac{A^3 a^3}{c}\left\{\frac{3\omega^2}{16\pi}\left(\frac{a}{c}\right)^2(n_2 + 6p_2)\right\}. \quad \text{(vii.)}$$

The expression for E_5 may be written down by symmetry. Collecting results from (v.), (vi.), and (vii.), we get, for the whole exhaustion of energy,

$$E \div \left(\frac{4\pi}{3}\right)^2 \frac{A^3 a^3}{c}$$
$$= \sum_{k=2}^{k=\infty}\left[\tfrac{3}{2}\left(\frac{A}{c}\right)^3 \frac{\gamma^{k-1}}{k-1}\left\{n_k - \tfrac{1}{2}n_k^2 - \frac{k+2!}{4.k-2!}p_k^2\right\} + \tfrac{3}{2}\left(\frac{a}{c}\right)^3\frac{\Gamma^{k-1}}{k-1}\left\{N_k - \tfrac{1}{2}N_k^2 - \frac{k+2!}{4.k-2!}P_k^2\right\}\right]$$
$$+ \left(\tfrac{3}{2}\right)^2\left(\frac{A}{c}\right)^3\left(\frac{a}{c}\right)^3 \sum_{k=2}^{k=\infty}\sum_{i=2}^{i=\infty}\frac{\Gamma^{i-1}}{i-1}\,\frac{\gamma^{k-1}}{k-1}\left\{\frac{k+i!}{k!\,i!}N_i n_k + \frac{k+2!}{2.k-2!}\,\frac{k+i!}{i-2!\,k+2!}P_i p_k\right\}$$
$$+ \frac{3\omega^2}{16\pi}\left[\left(\frac{a}{c}\right)^2(n_2 + 6p_2) + \left(\frac{A}{c}\right)^2(N_2 + 6P_2)\right]. \quad \text{(viii.)}$$

The expression is found without any assumption that the two masses are bounded by level surfaces, and therefore in equilibrium. But the condition for equilibrium is that the differential coefficients of E with respect to any one and all of the parameters n, p, N, P, shall vanish. If we equate to zero dV/dn_k, we get

$$1 - n_k + \tfrac{3}{2}\left(\frac{a}{c}\right)^3 \sum_{i=2}^{i=\infty}\frac{\Gamma^{i-1}}{i-1}\,\frac{k+i!}{k!\,i!}N_i = 0.$$

If, however, $k = 2$, there is on the left-hand side an additional term

$$\frac{3\omega^2}{16\pi}\left(\frac{a}{c}\right)^2 \div \tfrac{3}{2}\left(\frac{A}{c}\right)^3\frac{\gamma}{1} = \frac{\omega^2}{8\pi}\left(\frac{c}{A}\right)^3 = \tfrac{3}{16}e\left(\frac{c}{A}\right)^3.$$

The equation of dV/dN_i to zero gives a similar equation.

If we equate to zero dV/dp_k, we get

$$-p_k + \tfrac{3}{2}\left(\frac{a}{c}\right)^3 \Sigma\,\frac{\Gamma^{i-1}}{i-1}\,\frac{k+i!}{i-2!\,k+2!}P_i = 0.$$

If, however, $k = 2$, there is on the left-hand side an additional term

$$\frac{3\omega^2}{16\pi}\, 6\left(\frac{a}{c}\right)^3 + \tfrac{3}{2}\left(\frac{A}{c}\right)^3 \frac{\gamma}{1}\,\frac{4\,!}{2\,!\,0\,!} = \frac{\omega^2}{8\pi}\left(\frac{c}{A}\right)^3 \frac{6}{4\,.\,3} = \tfrac{1}{15}\,\epsilon\left(\frac{c}{A}\right)^3.$$

The equation of dV/dP_i to zero gives a similar equation.

Now, if we put $h_k + l_k$ for n_k, except when $k = 2$, and then put $h_2 + l_2 + \tfrac{2}{15}\epsilon\ (c/A)^3 = n_2$, and similarly introduce the H's and L's; and if we put $p_k = m_k$, except when $k = 2$, and then put $p_2 = m_2 + \tfrac{1}{15}\epsilon\,(c/A)^3$, and similarly introduce the M's, it is easy to see that the equations (i.) to the two surfaces become the same as (72), and the equations of condition between n and N, and between p and P, become exactly those which we found by a different method above in (23), (44), and (59). The only difference is that the equations for h and l are fused together.

This, therefore, forms a valuable confirmation of the correctness of the long analysis employed for the determination of the forms of equilibrium.

The formula (viii.) also enables us to obtain the intrinsic energy of the system, that is to say, the exhaustion of energy of the concentration of the matter from a state of infinite dispersion to its actual shape, with its sign changed.

The last line of (viii.) depends on the yielding of the fluid to centrifugal force, and must be omitted from the exhaustion of energy.

Besides the rest of (viii.), we have in the exhaustion of energy of the system, the exhaustion of the two spheres and their mutual exhaustion.

It is clear, then, that the intrinsic energy is

$$-\left(\tfrac{4}{3}\pi\right)^2 . \tfrac{3}{5}\left(a^5 + A^5\right) - \left(\tfrac{4}{3}\pi\right)^2 \frac{A^3 a^3}{c}$$

$$-\left(\tfrac{4}{3}\pi\right)^2 \frac{A^3 a^3}{c} \sum_{k=2}^{k=\infty}\left[\tfrac{3}{2}\left(\frac{A}{c}\right)^3 \frac{\gamma^{k-1}}{k-1}\left\{n_k - \tfrac{1}{2}n_k^2 - \frac{k+2\,!}{4\,.\,k-2\,!}\,p_k^2\right\} + \tfrac{3}{2}\left(\frac{a}{c}\right)^3 \frac{\Gamma^{k-1}}{k-1}\left\{N_k - \tfrac{1}{2}N_k^2 - \frac{k+2\,!}{4\,.\,k-2\,!}\,P_k^2\right\}\right]$$

$$-\left(\tfrac{4}{3}\pi\right)^2 \frac{A^3 a^3}{c} \sum_{k=2}^{k=\infty}\sum_{i=2}^{i=\infty}\left(\tfrac{3}{2}\right)^2 \left(\frac{A}{c}\right)^3 \left(\frac{a}{c}\right)^3 \frac{\Gamma^{i-1}}{i-1}\,\frac{\gamma^{k-1}}{k-1}\left\{\frac{k+i\,!}{k\,!\,i\,!}\,N_i n_k + \frac{k+i\,!}{2\,.\,k-2\,!\,i-2\,!}\,P_i p_k\right\}, \quad . \quad . \quad . \quad \text{(ix.)}$$

where the n, N, P, p, have their values determined in accordance with the condition that the surfaces are level surfaces.

In evaluating the intrinsic energy from this formula, it is convenient to refer the energy to that of a sphere of such radius, b, that its mass is equal to the whole mass of the system. Then $b^3 = a^3 + A^3$, and we may take the intrinsic energy as

$$\left(\tfrac{4}{3}\pi\right)^2 b^5 \left(i - 1\right).$$

Thus i will be a numerical quantity which is positive.

I find from (ix.), with $\beta = \tfrac{1}{5}$ and $a = A$, that $i = {\cdot}4873$.

With regard to the kinetic energy of the system, we have seen in § 10 that the moment of momentum is $\left(\tfrac{4}{3}\pi\right)^{\frac{3}{2}} b^5 \times \mu$, where μ is a numerical quantity, and we find in the course of the determination the function $3\omega^2/4\pi$. Then, since the energy is half the moment of momentum multiplied by the angular velocity, it is clear that the kinetic energy is

$$\left(\tfrac{4}{3}\pi\right)^2 b^5 \times \tfrac{1}{2}\mu \left(\frac{3\omega^2}{4\pi}\right).$$

The kinetic energy, as represented by $\epsilon = \tfrac{1}{2}\mu\left(3\omega^2/4\pi\right)^{\frac{1}{2}}$, is comparable with the intrinsic energy as represented by i.

In the case of $\beta = \tfrac{1}{5}$, and $a = A$, I find the kinetic energy $\epsilon = {\cdot}0925$.

Thus the total energy $\mathrm{E} = \epsilon + i = {\cdot}5798$.*

* [See foot-note added to the Summary above. October 10, 1887.]

If the energy of a system be expressed as the sum of a number of coefficients, each multiplied by the square of a parameter, it has been shown by M. POINCARÉ that the stability of the system depends on the signs of these coefficients, which he calls "the coefficients of stability." But, if the expression for the energy involves the products as well as the squares of the parameters, the coefficients of stability are the roots of a determinantal equation involving the second differentials of the energy with respect to the parameters.

Let V, a linear quadratic function of x, y, z, &c., be the energy of a system in equilibrium; then the determinantal equation is

$$\begin{vmatrix} \frac{d^2V}{dx^2} - \lambda, & \frac{d^2V}{dx\,dy}, & \frac{d^2V}{dx\,dz}, & \&c. \\ \frac{d^2V}{dy\,dx}, & \frac{d^2V}{dy^2} - \lambda, & \frac{d^2V}{dy\,dz}, & \&c. \\ \frac{d^2V}{dz\,dx}, & \frac{d^2V}{dz\,dy}, & \frac{d^2V}{dz^2} - \lambda, & \&c. \\ \&c., & \&c., & \&c. & \end{vmatrix} = 0. \quad \text{(x.)}$$

The solution of this equation in λ involves the determination of the several fundamental modes of vibration of the system; and the roots are the coefficients of stability.

Now suppose that V involves a constant, then, in causing that constant to vary continuously, we have a series of systems of equilibrium of the same kind; and the coefficients of stability vary continuously at the same time. If the system be initially in stable equilibrium, the stability ceases when a coefficient of equilibrium vanishes. The system at the moment of instability is in a condition of "bifurcation," that is to say, there is another series of shapes of a different kind, of which this shape is a member. In making the constant vary past the critical value, we find this second series of shapes stable, whilst the first is unstable.

If the system be in uniform rotation, so that instead of absolute equilibrium there is equilibrium relatively to uniformly rotating axes, the same theorems hold true, provided that only one root of the determinantal equation vanishes at a time.

This last is the case which we are considering, and the constant, which we suppose to vary continuously, is c, the distance between the two centres of the mean spheres of radii a and A.

When the two masses are far apart the equilibrium is stable, but when they are brought closer a time may come when one of the coefficients of stability vanishes.

The condition for the vanishing of a coefficient of stability is determined by the determinant (x.) with $\lambda = 0$.

To find the determinant, we have to evaluate the second differentials of E with respect to the parameters n, N, p, P.

If we form the determinant corresponding to (x.) with $\lambda = 0$, it is obvious that two infinite squares of entries which are diagonally opposite to one another, and which meet at a corner, are to be filled in with differentials involving $dn\,dp$, $dn\,dP$, $dN\,dP$, $dN\,dp$, in the denominators. All these entries are zero, and hence the infinite determinant splits into two independent infinite determinants, one only involving the differentials with respect to N, n, and the other only those with respect to P, p. The N, n determinant may be called "the tidal determinant," the P, p one "the rotational determinant"; for the origin of the terms in each is obvious.

By considering only the tidal determinant, we see how the other may be treated very shortly.

For the sake of brevity, write

$$n_k N_i = -\frac{d^2E/dn_k\,dN_i}{(-d^2E/dn_k^2)^{\frac{1}{2}}\,(-d^2E/dN_i^2)^{\frac{1}{2}}} \quad \text{(xi.)}$$

Then stability vanishes, as far as regards the tidal forces, when

$$\left|\begin{array}{cccccccc} \cdot & \cdot & \cdot & \cdot & \cdot & \cdot & \cdot & \cdot \\ \ldots, & 1, & 0, & 0, & n_4N_2, & n_4N_3, & n_4N_4, & \ldots \\ \ldots, & 0, & 1, & 0, & n_3N_2, & n_3N_3, & n_3N_4, & \ldots \\ \ldots, & 0, & 0, & 1, & n_2N_2, & n_2N_3, & n_2N_4, & \ldots \\ \ldots, & N_2n_4, & N_2n_3, & N_2n_2, & 1, & 0, & 0, & \ldots \\ \ldots, & N_3n_4, & N_3n_3, & N_3n_2, & 0, & 1, & 0, & \ldots \\ \ldots, & N_4n_4, & N_4n_3, & N_4n_2, & 0, & 0, & 1, & \ldots \\ \cdot & \cdot & \cdot & \cdot & \cdot & \cdot & \cdot & \cdot \end{array}\right| = 0. \quad \ldots \quad \text{(xii.)}$$

It will be observed that, by the notation (xi.), and the appropriate division of the columns and rows of the tidal determinant, it has been converted into a determinant in which the diagonal consists of ones.

If we drop certain factors common to the whole, we have, by differentiating (viii.),

$$\frac{d^2E}{dn_k^2} = -\tfrac{3}{2}\left(\frac{A}{c}\right)^3 \frac{\gamma^{k-1}}{k-1}, \quad \frac{d^2E}{dN_k^2} = -\tfrac{3}{2}\left(\frac{a}{c}\right)^3 \frac{\Gamma^{k-1}}{k-1}, \quad \frac{d^2E}{dn_k\,dn_i} = \frac{d^2E}{dN_k\,dN_i} = 0,$$

$$\frac{d^2E}{dn_k\,dN_i} = (\tfrac{3}{2})^2 \left(\frac{A}{c}\right)^3\left(\frac{a}{c}\right)^3 \frac{\gamma^{k-1}}{k-1}\,\frac{\Gamma^{i-1}}{i-1}\,\frac{k+i\,!}{k\,!\,i\,!},$$

$$\frac{d^2E}{dp_k^2} = -\tfrac{3}{2}\left(\frac{A}{c}\right)^3 \frac{k+2\,!}{2\,.\,k-2\,!}\,\frac{\gamma^{k-1}}{k-1}, \quad \frac{d^2E}{dP_k^2} = -\tfrac{3}{2}\left(\frac{a}{c}\right)^3 \frac{k+2\,!}{2\,.\,k-2\,!}\,\frac{\Gamma^{k-1}}{k-1}, \quad \frac{d^2E}{dp_k\,dp_i} = \frac{d^2E}{dP_k\,dP} = 0$$

$$\frac{d^2E}{dp_k\,dP_i} = (\tfrac{3}{2})^2 \left(\frac{A}{c}\right)^3\left(\frac{a}{c}\right)^3 \frac{k+2\,!}{2\,.\,k-2\,!}\,\frac{\gamma^{k-1}}{k-1}\cdot\frac{\Gamma^{i-1}}{i-1}\,\frac{k+i\,!}{i-2\,!\,k+2\,!},$$

$$= (\tfrac{3}{2})^2 \left(\frac{A}{c}\right)^3\left(\frac{a}{c}\right)^3 \tfrac{1}{2}\,\frac{\gamma^{k-1}}{k-1}\,\frac{\Gamma^{i-1}}{i-1}\,\frac{k+i\,!}{i-2\,!\,k-2\,!}; \quad \ldots \quad \text{(xiii.)}$$

$$\frac{d^2E}{dn_k\,dp_i} = \frac{d^2E}{dn_k\,dP_i} = \frac{d^2E}{dN_k\,dP_i} = \frac{d^2E}{dN_k\,dp_i} = 0.$$

With these values (xiii.) we easily find, by substitution in (xi.), that

$$n_kN_i = -\tfrac{3}{2}\left(\frac{A}{c}\right)^{\frac{3}{2}}\left(\frac{a}{c}\right)^{\frac{3}{2}}\left[\frac{\gamma^{k-1}}{k-1}\,\frac{\Gamma^{i-1}}{i-1}\right]^{\frac{1}{2}}\frac{k+i\,!}{k\,!\,i\,!}. \quad \ldots \quad \text{(xiv.)}$$

[This expression gives the value of each of the entries in the infinite determinant (xii.).

Now it is possible, by a certain laborious investigation which I do not here reproduce, to develope this infinite "tidal" determinant in the form of a series, and then to show that, however close the two masses may be to one another, the series arising from the tidal determinant can never vanish. It may also be proved that the other infinite determinant, which results from "rotational" terms, is necessarily greater than the tidal determinant, and *à fortiori* can never vanish.]*

Thus it might be held that stability must subsist in the figures of equilibrium until the two masses come into contact. But, as appears from § 11, it is certain that, if one of the masses be smaller than the other, this cannot be the case. In fact, the investigation must break down on account of the imperfection arising from the use of spherical harmonic analysis.

We have seen that the infinite determinant, which gives the coefficients of stability, splits into two parts when we rely on spherical harmonic analysis; but when instability ensues it must be brought about by the joint action of the tidal and rotational forces. It appears certain then that, if a rigorous analysis were used, this separation would not take place.

Although the present investigation proves thus to be abortive, I have thought it worth while to sketch the process, because it almost certainly indicates the line that must be pursued whenever a more rigorous analysis shall be applied to this difficult problem.

---

* This paragraph, replacing the investigation referred to, added July 12, 1887.

Dumb-bell figure of equilibrium.
Section perpendicular to axis of rotation.

$[A = a;\ \frac{c}{a} = 2 \cdot 449;\ \beta = \frac{1}{5};\ \gamma = \frac{1}{6};\ \frac{\omega^2}{4\pi} = \cdot 0494;\ \text{momentum} = (\tfrac{4}{3}\pi)^{\frac{4}{3}} b^5 \times \cdot 482]$

Fig. 2.

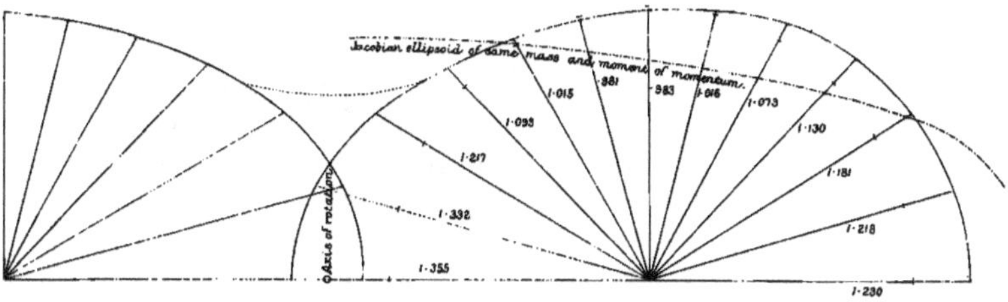

Dumb-bell figure of equilibrium
Section through axis of rotation

$[A = a;\ \frac{c}{a} = 2 \cdot 449;\ \beta = \frac{1}{5};\ \gamma = \frac{1}{6};\ \frac{\omega^2}{4\pi} = \cdot 0494;\ \text{momentum} = (\tfrac{4}{3}\pi)^{\frac{4}{3}} b^5 \times \cdot 482]$

Fig. 3.

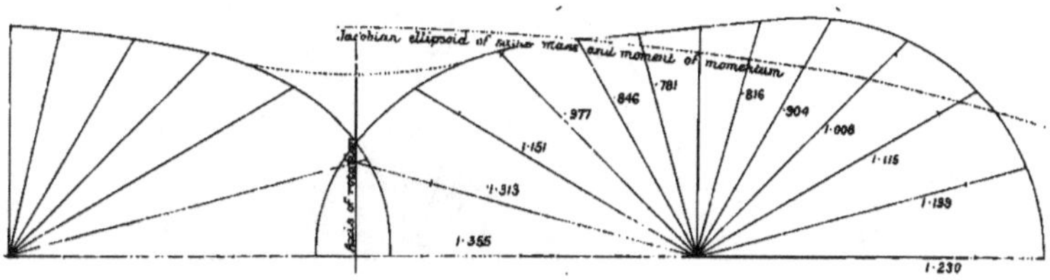

Equal masses nearly in contact.
Section perpendicular to axis of rotation

$[A = a;\ \frac{c}{a} = 2 \cdot 646;\ \beta = \frac{1}{6};\ \gamma = \frac{1}{7};\ \frac{\omega^2}{4\pi} = \cdot 038;\ \text{momentum} = (\tfrac{4}{3}\pi)^{\frac{4}{3}} b^5 \times \cdot 472]$

Fig. 4.

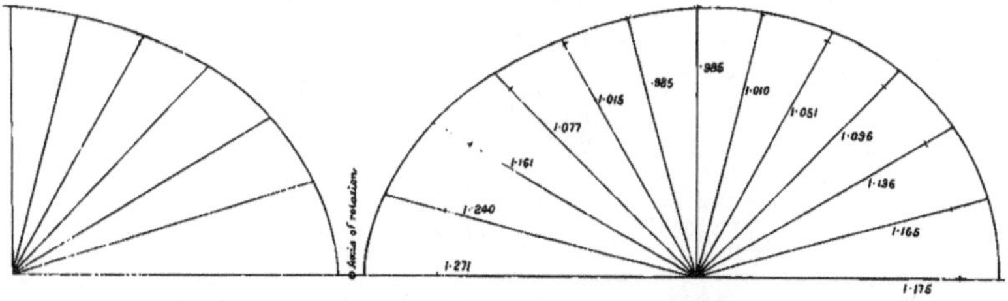

West, Newman & Co. lith.

Equal masses nearly in contact.
Section through axis of rotation.
$[A = a;\ \frac{c}{a} = \cdot 2046;\ \beta = \frac{1}{6};\ \gamma = \frac{1}{7};\ \frac{\omega^2}{4\pi} = \cdot 038;\ \text{momentum} = (\frac{4}{3}\pi)^{\frac{1}{2}} b^5 \times \cdot 472]$

Fig. 5.

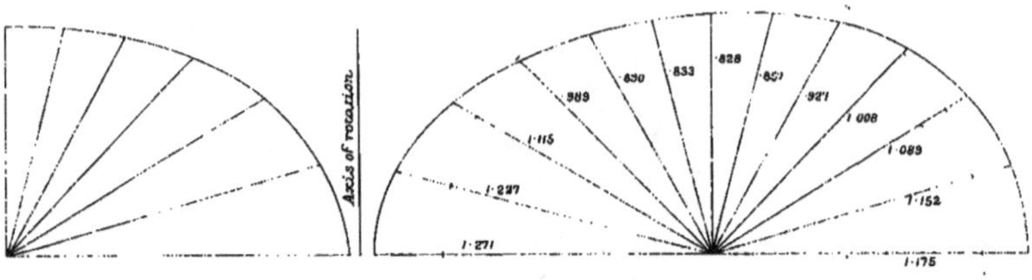

Unequal masses; section perpendicular to axis of rotation.
$[\frac{A}{a} = 3;\ (\frac{A}{a})^3 = 27;\ \frac{a}{c} = \cdot 189;\ \frac{A}{c} = \cdot 566;\ \gamma = \cdot 036;\ \Gamma = \cdot 321$
$\frac{\omega^2}{4\pi} = \cdot 066;\ \text{momentum} = (\frac{4\pi}{3})^{\frac{1}{2}} b^5 \times \cdot 29]$

Fig. 6.

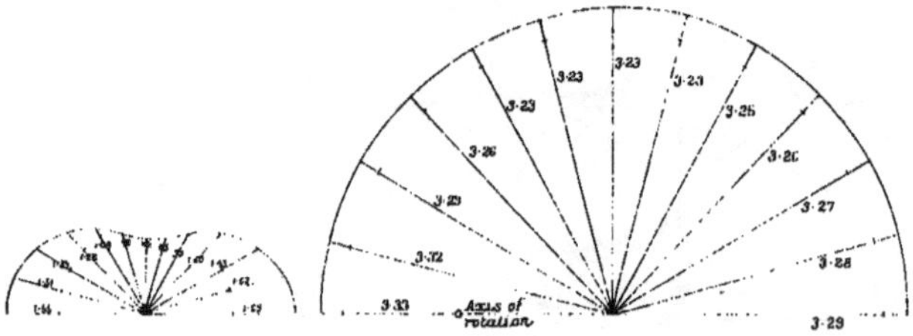

Unequal masses; section through axis of rotation.
$[\frac{A}{a} = 3;\ (\frac{A}{a})^3 = 27;\ \frac{a}{c} = \cdot 186;\ \frac{A}{c} = \cdot 566;\ \gamma = \cdot 036;\ \Gamma = \cdot 321$
$\frac{\omega^2}{4\pi} = \cdot 066;\ \text{momentum} = (\frac{4\pi}{3})^{\frac{1}{2}} b^5 \times \cdot 29]$

Fig. 7.

www.ingramcontent.com/pod-product-compliance
Lightning Source LLC
Chambersburg PA
CBHW022022190326
41519CB00010B/1570
9783337103644